Springer Monographs in Mathematics

Springer

London
Berlin
Heidelberg
New York
Barcelona
Hong Kong
Milan
Paris
Singapore
Tokyo

Wai Ki Ching

Iterative Methods for Queuing and Manufacturing Systems

With 17 Figures

 Springer

Wai Ki Ching
Faculty of Mathematical Studies
University of Southampton
Highfield
Southampton
SO17 1BJ
UK

Springer Monographs in Mathematics ISSN 1439-7382

ISBN 1-85233-416-9 Springer-Verlag London Berlin Heidelberg

British Library Cataloguing in Publication Data
Ching, Wai Ki
  Iterative methods for queuing and manufacturing systems. -
  (Springer monographs in mathematics)
  1. Markov processes 2. Iterative methods (Mathematics)
  3. Queuing Theory 4. System analysis 5. Flexible manufacturing
  systems – Mathematical models
  I. Title
  519
  ISBN 1852334169

Library of Congress Cataloging-in-Publication Data
Ching, Wai Ki, 1969-
Iterative methods for queuing and manufacturing systems / Wai Ki Ching.
    p. cm. -- (Springer monographs in mathematics)
  Includes bibliographical references.
  ISBN 1-85233-416-9 (alk. paper)
  1. Queuing Theory 2. Iterative methods (Mathematics) I. Title. II. Series.
T57.9.C48 2001
519.82—dc21   00-052659

Mathematics Subject Classification (1991): 65F10, 60J22, 90B05, 90B22

© Springer-Verlag London Limited 2001
Printed in Great Britain

Typesetting: Camera-ready by the author
Printed and bound at the Athenæum Press Ltd., Gateshead, Tyne & Wear
12/3830-543210 Printed on acid-free paper SPIN 10790110

To Mandy Lee, Shun Tai Ching,
and Kam Koo Wong

Preface

The aim of this book is to outline the recent development of iterative methods for solving Markovian queuing and manufacturing systems. Markov processes are widely used in the modeling of queuing systems, manufacturing systems, telecommunication systems, computer systems, and many other practical systems. Very often, in the system performance analysis, one faces the problem of solving the system steady-state probability distribution of a large number of states. Classical iterative methods such as Gauss-Seidel (GS) and Jacobi methods are the common iterative methods used to tackle the problem. However, their convergence rates are slow in general and very often increase linearly with respect to the number of states in the system. In this book, we consider Conjugate Gradient (CG) type methods in solving the steady-state probability distribution of queuing and manufacturing systems. Usually, the CG method is used with a matrix, called a preconditioner, to accelerate its convergence rate. We construct preconditioners for Markovian queuing and manufacturing systems such that the Preconditioned Conjugate Gradient (PCG) method converges very fast. The construction techniques can be applied to a large class of practical systems. Apart from computational methods, interesting Markovian models with applications in inventory control and supply chains are also introduced.

This book consists of nine chapters. Chapter 1 gives a brief introduction, with overviews of Markov processes, simple queuing and manufacturing systems, and iterative methods.

In Chapter 2, we discuss Toeplitz-circulant preconditioners for queuing systems with batch arrivals. In Chapter 3, we present a queuing system with Markov Modulated Poisson Process (MMPP) inputs, which arise in telecommunication systems. The CG method is applied to solve the steady-state probability distribution with the preconditioner constructed by taking a circulant approximation of the generator matrix.

In Chapter 4, we discuss an application of MMPP to the manufacturing systems of multiple failure-prone machines under Hedging Point Production (HPP) policy. In Chapter 5, we consider failure-prone manufacturing systems with batch arrivals of demand. A preconditioner is constructed by exploiting the structure of the generator matrix of the system.

In Chapter 6, we model a flexible manufacturing system by using a Markovian queue. In Chapter 7, we discuss a manufacturing system of two machines in tandem. HPP policy is employed as the production control. The steady-state probability distribution of the system is then solved by the CG method with a preconditioner constructed by taking circulant approximation of the generator matrix.

In Chapter 8, we discuss an analytical model for manufacturing systems under delivery time guarantee policy. An analytical solution of the steady-state probability distribution of the inventory system is obtained and is applied to the analysis of the model. In Chapter 9, we discuss an application of MMPP in modeling multi-location inventory systems. A Markovian queuing model for supply chains problems is also introduced.

This book is aimed at students, professionals, practitioners, and researchers in applied mathematics, scientific computing, and operational research, who are interested in the formulation and computation of queuing and manufacturing systems. Readers are expected to have some basic knowledge of Markov processes and matrix theory.

Acknowledgements

It is a pleasure to thank the following people and organizations. The research described herein is supported by the Croucher Foundation. I am indebted to many former and present colleagues who collaborated on the ideas described here. I would like to thank Prof. Raymond H. Chan and Prof. X.Y. Zhou for introducing me to the subjects of queuing networks and manufacturing systems, and for their guidance and help throughout the years. I would also like to thank Prof. T.M. Shih, Dr. S. Scholtes, and an anonymous reviewer for their helpful encouragement and comments; without them this book would not have been possible.

All the figures, tables, propositions and lemmas in Chapters 2, 3 and 4 are reprinted from [25, 50] with permission from SIAM. All the figures, tables, propositions and lemmas in Chapters 5, 7, 8 and 9 are reprinted from [35, 36, 40, 41] with the permission from Elsevier Science.

Southampton, December 2000   *Wai Ki CHING*

List of Figures

Contents

1. Introduction and Overview

1.1 Introduction

In this book, we discuss iterative methods and Markovian models for queuing and manufacturing systems. We are interested in obtaining the steady-state probability distributions of these systems because much important system performance information such as the blocking probability, throughput, and average running cost of the system can be written in terms of this probability distribution. In some simple situations, it is possible to derive the analytical or an approximated steady-state probability distribution for a queuing or manufacturing system. But very often solving the steady-state probability distribution is a difficult task. Iterative methods and simulation techniques are common and powerful tools for solving the captured problems. The main focus of this book is the application of iterative methods in solving Markovian queuing and limiting probability systems. For simulation methods, we refer interested readers to the textbooks of references [86, 93, 95].

In this chapter, we expect to give readers brief introductions to Markov processes, CG methods and simple queuing systems. The chapter is organized as follows. An introduction to Markov processes is given in Section 1.2. Some simple examples of Markovian queuing systems are given in Section 1.3. An introduction to iterative methods and the Conjugate Gradient (CG) method is given in Section 1.4. Finally, a summary of the remaining chapters in this book is given in Section 1.5.

1.2 Markov Process

A *Markov process* (or *Markov chain*) is a probabilistic dynamic system of states in which the future state depends only on the present situation and is independent of the past history. The states of a Markov process may be discrete or continuous. In our discussion, we study discrete models of queuing and manufacturing systems. Markov processes can also be classified into two categories: *discrete time* and *continuous time*. Let us look at a simple example of a discrete time Markov process of two states.

1.2.1 Discrete Time Markov Process of Two States

Consider a machine which consists of two states: down (0) and normal (1). Let $x(t)$ ($x(t) = 1$ or $x(t) = 0$) be the state at time $t(t = 0, 1, \ldots,)$ and $p(x(t), t)$ be the probability that the machine is in state $x(t)$ at time t. We assume that at the beginning the machine is normal, i.e. $x(0) = 1$, and therefore $p(1, 0) = 1$ and $p(0, 0) = 0$. Suppose that the system dynamic is described by the following conditional probabilities. Let r be the conditional probability that the system is in state 0 at time $t + 1$, given that it is in state 1 at time t. Let q be the conditional probability that the machine is in state 1 at time $t + 1$, given that it is in state 0 at time t. One may write

$$\text{Prob}(x(t + 1) = 0 | x(t) = 1) = r \quad \text{and} \quad \text{Prob}(x(t + 1) = 1 | x(t) = 0) = q. \tag{1.1}$$

Since at time $t + 1$, the machine can be either normal or down, we have also

$$\text{Prob}(x(t+1) = 1 | x(t) = 1) = 1 - r \quad \text{and} \quad \text{Prob}(x(t+1) = 0 | x(t) = 0) = 1 - q. \tag{1.2}$$

This is a Markov process, because the future behavior of the machine at time $t + 1$ depends on the current state at time t only and independent of its past history. We are interested in the probabilities $p(1, t)$ and $p(0, t)$. By the above system dynamic, one may write down the following equations:

$$\begin{cases} p(0, t + 1) = (1 - q)p(0, t) + rp(1, t) \\ p(1, t + 1) = qp(0, t) + (1 - r)p(1, t) \end{cases} \tag{1.3}$$

for $t = 0, 1, 2, \ldots$. In matrix terms, we may write

$$
\begin{aligned}
\begin{pmatrix} p(0, t + 1) \\ p(1, t + 1) \end{pmatrix} &= \begin{pmatrix} 1 - q & r \\ q & 1 - r \end{pmatrix} \begin{pmatrix} p(0, t) \\ p(1, t) \end{pmatrix} \\
&= \begin{pmatrix} 1 - q & r \\ q & 1 - r \end{pmatrix} \begin{pmatrix} 1 - q & r \\ q & 1 - r \end{pmatrix} \begin{pmatrix} p(0, t - 1) \\ p(1, t - 1) \end{pmatrix} \\
&= \begin{pmatrix} 1 - q & r \\ q & 1 - r \end{pmatrix}^2 \begin{pmatrix} p(0, t - 1) \\ p(1, t - 1) \end{pmatrix} \\
&= \begin{pmatrix} 1 - q & r \\ q & 1 - r \end{pmatrix}^{t+1} \begin{pmatrix} p(0, 0) \\ p(1, 0) \end{pmatrix}.
\end{aligned} \tag{1.4}
$$

Proposition 1.2.1. *The solution of the difference equations (1.4) can be shown to be (see Exercise 1.2)*

$$\begin{cases} p(0, t) = p(0, 0)(1 - q - r)^t + \dfrac{r}{q + r}[1 - (1 - r - q)^t] \\ p(1, t) = p(1, 0)(1 - q - r)^t + \dfrac{q}{q + r}[1 - (1 - r - q)^t]. \end{cases} \tag{1.5}$$

An important and interesting thing to note is the long-term behavior of the system, i.e.

$$p_0 = \lim_{t \to \infty} p(0, t) \quad \text{and} \quad p_1 = \lim_{t \to \infty} p(1, t). \tag{1.6}$$

From Proposition 1.2.1, it is clear that if q and r do not take extreme values, then

$$p_0 = \lim_{t \to \infty} p(0, t) = \frac{r}{q + r} \quad \text{and} \quad p_1 = \lim_{t \to \infty} p(1, t) = \frac{q}{q + r}. \tag{1.7}$$

In fact, one can obtain the *limiting probabilities* or the *steady-state probabilities* p_0 and p_1 without solving (1.4). Suppose the steady-state probability distribution exists, we let t go to infinity in (1.4) and get

$$\begin{pmatrix} p_0 \\ p_1 \end{pmatrix} = \begin{pmatrix} 1 - q & r \\ q & 1 - r \end{pmatrix} \begin{pmatrix} p_0 \\ p_1 \end{pmatrix}. \tag{1.8}$$

Hence $(p_0, p_1)^t$ is the solution of the homogeneous linear equations

$$\begin{pmatrix} q & -r \\ -q & r \end{pmatrix} \begin{pmatrix} p_0 \\ p_1 \end{pmatrix} \equiv A \begin{pmatrix} p_0 \\ p_1 \end{pmatrix} = \begin{pmatrix} 0 \\ 0 \end{pmatrix}. \tag{1.9}$$

The matrix A is called the *generator matrix* of the system. We note that A is singular and therefore the solution for (1.9) is not unique. To obtain a unique solution, one must make use of the fact that $(p_0, p_1)^t$ is a probability distribution, i.e.

$$p_0 + p_1 = 1.$$

One sufficient condition for the existence of steady-state probability in a Markov process of finite states is that the Markov chain itself is *irreducible*; see Varga [113] for instance. A Markov chain is said to be irreducible if its underlying graph has the following property: for any two vertices (states) i and j there is a directed path from $i(j)$ to $j(i)$. This means there is a non-zero probability from state $i(j)$ to state $j(i)$. For example the Markov chain of the two-state process (see Fig 1.1) is irreducible. The meaning of steady state is

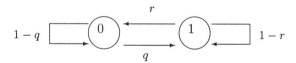

Fig. 1.1. The Markov chain for the two-state system.

that the performance measures appear to be close to constant as time goes to infinity. In the above example, p_1 is the proportion of time that the machine is normal in the long run and p_0 is the proportion of time that the machine is down in the long run.

1.2.2 Continuous Time Markov Process

In many practical systems, a change of state does not occur at a fixed discrete time. In fact, very often the duration of a system state is a continuous random variable. In our context, we model the system dynamic by the *Poisson process*. A process is called a Poisson process if

(A1) the probability of occurrence of one event in the time interval $(t, t + \delta t)$ is $\lambda \delta t + o(\delta t)$. Here λ is a positive constant and $o(\delta t)$ is such that

$$\lim_{\delta t \to 0} \frac{o(\delta t)}{\delta t} = 0.$$

(A2) the probability of occurrence of no event in the time interval $(t, t + \delta t)$ is $1 - \lambda \delta t + o(\delta t)$.

(A3) the probability of occurrences of more than one event is $o(\delta t)$.

Here an "event" can be an arrival of a customer or a departure of customer.

In the following, we are going to derive the well-known *Poisson distribution* and its relationship to another well known continuous distribution, the *exponential distribution*. They are important elements in Markovian queuing and manufacturing systems.

Let $P_n(t)$ be the probability that n events occurred in the time interval $[0, t]$. Suppose that $P_n(t)$ is differentiable. We are going to get a relationship between $P_n(t)$ and $P_{n-1}(t)$. Consider $P_n(t + \delta t)$ which is the probability of getting n events in the time interval $[0, t + \delta t]$. We can divide the time interval as the following two time intervals: $[0, t]$ and $(t, t + \delta]$. We know that in the time interval $(t, t + \delta]$, one event occurs, no event occurs, or more than one event occurs, and their probabilities are

$$\lambda \delta t + o(\delta t), \quad 1 - (\lambda \delta t + o(\delta t)), \quad \text{and} \quad o(\delta t)$$

respectively. Thus to have n events occurring in the time interval $[t, t + \delta t]$, we may have the following three cases. The first case is that we have n events occurring in the time interval $[0, t]$ and no more event occurring in the time interval $(t, t + \delta t]$; the probability is

$$P_n(t) \cdot (1 - \lambda \delta t - o(\delta t)). \tag{1.10}$$

The second case is that we have $(n - 1)$ events occurring in the time interval $[0, t]$ and one event occurring in the time interval $(t, t + \delta t]$; the probability is

$$P_{n-1}(t) \cdot (\lambda \delta t + o(\delta t)). \tag{1.11}$$

The third case is that we have less than $(n - 1)$ events occurring in the time interval $[0, t]$ and more than one event occurring in $(t, t + \delta t]$; the probability is $o(\delta t)$. From (1.10) and (1.11), we get

$$P_n(t + \delta t) = P_n(t) \cdot (1 - \lambda\delta t - o(\delta t)) + P_{n-1}(t) \cdot (\lambda\delta t + o(\delta t)) + o(\delta t). \quad (1.12)$$

Rearranging the terms in (1.12) we have

$$\frac{P_n(t + \delta t) - P_n(t)}{\delta t} = -\lambda P_n(t) + \lambda P_{n-1}(t) + (P_{n-1}(t) + P_n(t))\frac{o(\delta t)}{\delta t}. \quad (1.13)$$

Let $\delta t \to 0$ we have

$$\lim_{\delta t \to 0} \frac{P_n(t + \delta t) - P_n(t)}{\delta t} = -\lambda P_n(t) + \lambda P_{n-1}(t) + \lim_{\delta t \to 0} (P_{n-1}(t) + P_n(t))\frac{o(\delta t)}{\delta t}. \quad (1.14)$$

Hence

$$\frac{dP_n(t)}{dt} = -\lambda P_n(t) + \lambda P_{n-1}(t) + 0, \quad n = 0, 1, 2, \ldots. \quad (1.15)$$

Since $P_{-1}(t) = 0$, we have the *initial value problem* for $P_0(t)$ as follows:

$$\frac{dP_0(t)}{dt} = -\lambda P_0(t) \quad \text{with} \quad P_0(0) = 1. \quad (1.16)$$

The probability $P_0(0)$ is the probability that no event occurred in the time interval $[0, 0]$, so it must be one. Solving the separable ordinary differential equation for $P_0(t)$ we get

$$P_0(t) = e^{-\lambda t} \quad (1.17)$$

which is the probability that no event occurred in the time interval $[0, t]$. Thus

$$1 - P_0(t) = 1 - e^{-\lambda t}$$

is the probability that at least one event occurred in the time interval $[0, t]$ or equivalently the first event occurred in the time interval $[0, t]$. Thus the probability density function $f(t)$ for waiting time of the first event to occur is given by

$$f(t) = \frac{d(1 - e^{-\lambda t})}{dt} = \lambda e^{-\lambda t}, \quad t \geq 0. \quad (1.18)$$

This is the well-known exponential distribution and it is a continuous probability distribution.

Now let us go back to solve $P_n(t)$. We note that

$$\begin{cases} \dfrac{dP_n(t)}{dt} = -\lambda P_n(t) + \lambda P_{n-1}(t), & n = 1, 2, \ldots \\ P_0(t) = e^{-\lambda t}, \\ P_n(0) = 0 & n = 1, 2, \ldots. \end{cases} \quad (1.19)$$

When $n = 0$ we have $P_0(t) = e^{-\lambda t}$. When $n = 1$ we have

$$\frac{dP_1(t)}{dt} = -\lambda P_1(t) + \lambda P_0(t), \quad \text{with} \quad P_1(0) = 0. \quad (1.20)$$

Solving (1.20) we get

$$P_1(t) = \frac{\lambda t}{1!} e^{-\lambda t}. \tag{1.21}$$

When $n = 2$ we have

$$\frac{dP_2(t)}{dt} = -\lambda P_2(t) + \lambda P_1(t), \quad \text{with } P_2(0) = 0. \tag{1.22}$$

Solving (1.22) we get

$$P_2(t) = \frac{(\lambda t)^2}{2!} e^{-\lambda t}. \tag{1.23}$$

Inductively we guess

$$P_n(t) = \frac{(\lambda t)^n}{n!} e^{-\lambda t}. \tag{1.24}$$

By mathematical induction on n one can prove the following proposition and this is left as an exercise.

Proposition 1.2.2. *The solution for (1.19) is (1.24).*

We note that when n is fixed, $P_n(t)$ is a continuous distribution in t, but when $t(t > 0)$ is fixed, we obtain the Poisson distribution $P_n(t)$ which is a discrete distribution in $n(n = 0, 1, 2, \ldots)$. Thus in a Poisson process, the inter-occurrence time between two successive event is a exponentially distributed continuous random variable. The following is an important property of the exponential distribution. One can also obtain the following interesting equivalent relation.

Proposition 1.2.3. *The following statements (A),(B), and (C) are equivalent.*
(A) The arrival process is a Poisson process with parameter λ.
(B) Let $N(t)$ be the number of arrivals in the time interval $[0, t]$ and

$$\text{Prob}(N(t) = n) = \frac{(\lambda t)^n e^{-\lambda t}}{n!} \quad n = 0, 1, 2, \ldots.$$

(C) The inter-arrival time follows the exponential distribution with parameter λ.

Proof. From the discussion above we have already proved (A) $\Rightarrow$ (B) and (B) $\Rightarrow$ (C). We need only to prove (C) $\Rightarrow$ (A). We first observe that

$$\text{Prob}(N(t) = 0) = e^{-\lambda t} = 1 - \lambda t + o(t).$$

This is required by (A2) of a Poisson process.

Next we are going to find $\text{Prob}(N(t) = 1)$. Let T_1 and T_2 be the waiting time for the first arrival and the second arrival respectively. The event that $N(t) = 1$ occurs if the first arrival in the interval $[0, t]$ arrives in the dt interval

$[u, u + dt)$ for some $u < t$ and then there is no further arrival in the remaining period $[u + dt, t]$. This corresponds to the situation that

$$u < T_1 < u + dt \quad \text{and} \quad T_2 > t - u.$$

We have

$$\text{Prob}(u < T_1 < u + dt) = \lambda e^{-\lambda u}$$

and

$$\text{Prob}(T_2 > t - u) = e^{-\lambda(t-u)}.$$

Therefore

$$
\begin{aligned}
\text{Prob}(N(t) = 1) &= \int_0^t \text{Prob}(u < T_1 < u + dt)\text{Prob}(T_2 > t - u)du \\
&= \int_0^t \lambda e^{-\lambda u} e^{-\lambda(t-u)} du \\
&= \lambda t e^{-\lambda t} \\
&= \lambda t + o(t).
\end{aligned}
$$

This is required by (A1). We have also

$$
\begin{aligned}
\text{Prob}(N(t) > 1) &= 1 - \text{Prob}(N(t) = 1) - \text{Prob}(N(t) = 0) \\
&= 1 - (1 - \lambda t + o(t)) - (\lambda t + o(t)) \qquad (1.25) \\
&= o(t).
\end{aligned}
$$

This is required by (A3) and we complete our proof. □

A probability density function satisfying (1.26) in the following proposition is said to have *no-memory property*. This property implies that if one wants to know the probability distribution of the time until the next arrival, then it does not matter how long it has been since the last arrival.

Proposition 1.2.4. *If x follows the exponential distribution with parameter λ, then for all non-negative values of t and h, we have*

$$\text{Prob}(x > t + h | x \geq t) = \text{Prob}(x > h). \qquad (1.26)$$

Proof. It is clear that

$$\text{Prob}(x > h) = \int_h^\infty \lambda e^{-\lambda t} dt = e^{-\lambda h}.$$

We note that

$$\text{Prob}(x > h | x \geq t) = \frac{\text{Prob}(x > h + t \text{ and } x \geq t)}{\text{Prob}(x \geq t)}.$$

From (1.26) we get

$$\text{Prob}(x > h + t \text{ and } x \geq t) = e^{-\lambda(t+h)} \quad \text{and} \quad \text{Prob}(x \geq t) = e^{-\lambda t}.$$

Therefore

$$\text{Prob}(x > t + h | x \geq t) = \frac{e^{-\lambda(t+h)}}{e^{-\lambda t}} = \text{Prob}(x > h).$$

$\square$

It can be shown that there is no other continuous probability density function can satisfy (1.26); see Feller [66]. We close the discussion on exponential distribution with the following brief discussion of an interesting paradox, the *waiting time* paradox. Suppose the inter-arrival time of customers to a server is exponentially distributed with a mean of 1 h. If you arrive at the server at a randomly chosen time then what is the average waiting time for a customer? The no-memory property of the exponential distribution suggests that no matter how long it has been since the last customer arrived, the average waiting time for the next customer is still 1 h. This indeed is true. But one may argue in this way: on average, somebody who arrives at a random time should arrive in the middle of a typical interval, and the average waiting time between successive customers is 1 h, then one has to wait, on average 0.5 h for the next customer. Why is the argument wrong? It is because the typical interval between two customers is longer than 0.5 h. It is more likely to arrive during a longer interval than a shorter interval. Let us consider the following simple example. We assume that half of the inter-arrival times are 0.5 h (short inter-arrival time) and half of them are 1.5 h (long inter-arrival time). We further assume that a long inter-arrival time comes after a short inter-arrival time and vice versa. Obviously, the probability that one will arrive during a long inter-arrival time is 0.75 h and 0.25 h that one will arrive during a short inter-arrival time. The expected inter-arrival time within which a customer arrives is

$$\frac{3}{4}(1.5) + \frac{1}{4}(0.5) = 1.25 \text{ h}.$$

Since we do arrive in the middle of an inter-arrival time, the average waiting time will be $0.5(5/4) = 0.625$ h.

1.2.3 A One-Server System

Consider a one-server system of two states: 0 (idle) and 1 (busy). The arrival of customers follows a Poisson process with parameter λ. The service time follows the exponential distribution with mean time being μ^{-1}. There is no waiting space in the system. A customer will leave the system when he finds the server is busy. Let $P_0(t)$ be the probability that the server is idle at time t and $P_1(t)$ be the probability that the server is busy at time t. Using a similar argument in the derivation of Poisson processes we have

$$\begin{cases} P_0(t+\delta t) = (1 - \lambda \delta t - o(\delta t))P_0(t) + (\mu \delta t + o(\delta t))P_1(t) + o(\delta t) \\ P_1(t+\delta t) = (1 - \mu \delta t - o(\delta t))P_1(t) + (\lambda \delta t + o(\delta t))P_0(t) + o(\delta t). \end{cases} \quad (1.27)$$

In this one-server system (a queuing system), λ is called the *arrival rate* of the customer and μ is called the *service rate* of the system. Rearranging the terms in (1.27), we have

$$\begin{cases} \dfrac{P_0(t+\delta t) - P_0(t)}{\delta t} = -\lambda P_0(t) + \mu P_1(t) + (P_1(t) - P_0(t))\dfrac{o(\delta t)}{\delta t} \\ \dfrac{P_1(t+\delta t) - P_1(t)}{\delta t} = \lambda P_0(t) - \mu P_1(t) + (P_0(t) - P_1(t))\dfrac{o(\delta t)}{\delta t}. \end{cases} \quad (1.28)$$

Letting $\delta t \to 0$ in (1.28) we have

$$\begin{cases} \dfrac{dP_0(t)}{dt} = -\lambda P_0(t) + \mu P_1(t) \\ \dfrac{dP_1(t)}{dt} = \lambda P_0(t) - \mu P_1(t). \end{cases} \quad (1.29)$$

Adding the two equations in (1.29) we have

$$\frac{d(P_0(t) + P_1(t))}{dt} = 0, \quad (1.30)$$

and therefore

$$P_0(t) + P_1(t) = c. \quad (1.31)$$

Since at any time the server is either idle or busy, we must have

$$P_0(t) + P_1(t) = 1,$$

and therefore

$$P_0(t) = 1 - P_1(t).$$

Substituting this back into (1.29) we get

$$\frac{P_1(t)}{dt} = \lambda(1 - P_1(t)) - \mu P_1(t) = \lambda - (\lambda + \mu)P_1(t). \quad (1.32)$$

Solving (1.32) we get

$$P_1(t) = \alpha e^{-(\lambda+\mu)t} + \frac{\lambda}{\lambda + \mu}. \quad (1.33)$$

Since $P_1(0) = 1$, we get

$$\alpha = \frac{\mu}{\lambda + \mu}.$$

Therefore we have

$$P_1(t) = \frac{1}{\lambda + \mu}(\mu e^{-(\lambda+\mu)t} + \lambda) \quad (1.34)$$

and
$$P_0(t) = 1 - P_1(t).$$

It is interesting to note the steady-state probabilities:

$$\begin{cases} \lim_{t \to \infty} P_0(t) = \dfrac{\mu}{\lambda + \mu} \\ \lim_{t \to \infty} P_1(t) = \dfrac{\lambda}{\lambda + \mu}. \end{cases} \tag{1.35}$$

Again the steady-state probability distribution in (1.35) can be obtained without solving the differential equations (1.29). In fact, (1.29) can be written in matrix form:

$$\begin{pmatrix} \frac{dP_0(t)}{dt} \\ \frac{dP_1(t)}{dt} \end{pmatrix} = \begin{pmatrix} -\lambda & \mu \\ \lambda & -\mu \end{pmatrix} \begin{pmatrix} P_0(t) \\ P_1(t) \end{pmatrix}. \tag{1.36}$$

In steady state, $P_0(t) = p_0$ and $P_1(t) = p_1$ are constants and independent of t. Therefore we have

$$\frac{dp_0(t)}{dt} = \frac{dp_1(t)}{dt} = 0.$$

The steady-state probabilities will be the solution of the following linear system:

$$\begin{pmatrix} \lambda & -\mu \\ -\lambda & \mu \end{pmatrix} \begin{pmatrix} p_0 \\ p_1 \end{pmatrix} = \begin{pmatrix} 0 \\ 0 \end{pmatrix} \tag{1.37}$$

subject to $p_0 + p_1 = 1$.

So far in our examples, the discrete and continuous time Markov chains discussed have only two states. Therefore without much difficulty, one can obtain the system steady-state probability distribution in analytical form. However, in many practical situations, the number of states can be thousands and system dynamics can be much more complicated. Hence numerical methods such as iterative methods are designed to solve the problem.

We remark that in matrix terms, we are interested in obtaining the normalized right hand side null vectors of the generator matrices of the queuing systems. Conventionally, a generator matrix Q has non-negative off-diagonal entries and zero row sum. For the sake of presentation, in the discussion of the chapters, all the generator matrices are assume to take the form of $-Q^t$ which has non-positive off-diagonal entries and zero column sum. The queuing and manufacturing models considered in this book are irreducible Markov chains and therefore the existence of steady-state probability distribution is guaranteed; see Varga [113] for instance. Readers who are interested to know more about Markov processes may consult the books by Ross [99, 100].

1.3 Some Examples in Queuing and Manufacturing Systems

In this section we are going to introduce some simple Markovian queuing and manufacturing systems. For the captured systems we either derive their analytical steady-state probability distributions or we give an approximated solution for the steady-state probability distribution.

1.3.1 An M/M/s/$n - s - 1$ Queuing System

The one-server system discussed in Section 1.2.3 is a queuing system without waiting space. This means when a customer arrives and finds the server is busy, the customer has to leave the system. Now let us consider a more general queuing system with customer arrival rate being λ. Suppose the system has s parallel and identical exponential servers with service rate being μ and there are $n - s - 1$ waiting spaces in the system. The queuing discipline is *First-come-first-served*. When a customer arrives and finds all the servers are busy, the customer can still wait in the queue provided that there is a waiting space available. Otherwise, the customer has to leave the system; see Fig 1.2.

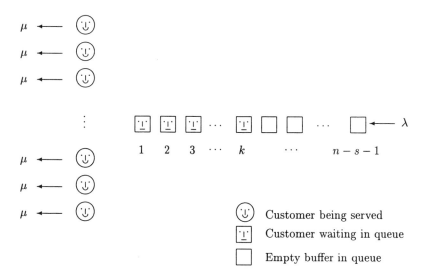

Fig. 1.2. The one-queue system.

To describe the queuing system, we use the number of customers in the queue to represent the state of the system. There are n states, namely $0, 1, \ldots, n - 1$. The Markov chain for the queuing system is given in Fig. 1.3. The number of customers in the system is used to represent the states in the Markov chain. Clearly it is an irreducible Markov chain.

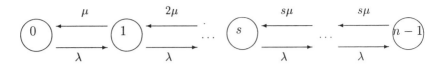

Fig. 1.3. The Markov chain for the one-queue system.

If we order the states of the system in increasing number of customers, it is not difficult to show that the generator matrix for this queuing system is given by the following $n \times n$ tri-diagonal matrix $A_1 = A_{(n,s,\lambda,\mu)}$ where

$$
A_1 = \begin{pmatrix}
\lambda & -\mu & & & & & & 0 \\
-\lambda & \lambda+\mu & -2\mu & & & & & \\
& \ddots & \ddots & \ddots & & & & \\
& & -\lambda & \lambda+(s-1)\mu & -s\mu & & & \\
& & & -\lambda & \lambda+s\mu & -s\mu & & \\
& & & & \ddots & \ddots & \ddots & \\
& & & & & -\lambda & \lambda+s\mu & -s\mu \\
0 & & & & & & -\lambda & s\mu
\end{pmatrix} \quad (1.38)
$$

and the underlying Markov chain is irreducible. The solution for the steady-state probability distribution can be shown to be (see Exercise 1.5)

$$
\mathbf{p}^t_{(n,s,\lambda,\mu)} = (p_0, p_1, \ldots, p_{n-1})^t \quad (1.39)
$$

where

$$
p_i = \alpha \prod_{k=1}^{i+1} \frac{\lambda}{\mu \min\{k,s\}} \quad \text{and} \quad \alpha^{-1} = \sum_{i=0}^{n} p_i. \quad (1.40)
$$

Here p_i is the probability that there are i customers in the queuing system in the steady state and α is the *normalization constant*.

Before we end this section, let us introduce the famous *Little's queuing formula*. If we define the following:

(a) λ, average number of arrivals entering the system,
(b) L_c, average number of customers in the queuing system,
(c) L_q, average number of customers waiting in the queue,
(d) L_s, average number of customers in service,
(e) W_c, average time a customer spends in the queuing system,
(f) W_q, average time a customer spends in waiting in the queue,
(g) W_s, average time a customer spends in service,

then Little's queuing formula states that if the steady-state probability distribution exists, we have

$$L_c = \lambda W_c, \quad L_q = \lambda W_q, \quad \text{and} \quad L_s = \lambda W_s.$$

Little's queuing formula is valid for a general queuing system and its proof can be found in Ross [98].

Example 1.3.1. Consider a one-server system; the steady-state probability distribution is given by

$$p_i = \frac{\rho^i(1-\rho)}{1-\rho^n} \quad \text{where} \quad \rho = \frac{\lambda}{\mu}.$$

When the system has no limit on waiting space and $\rho < 1$, the steady-state probability becomes

$$\lim_{n \to \infty} p_i = \rho^i(1-\rho).$$

The expected number of customers in the system is given by

$$
\begin{aligned}
L_c &= \sum_{i=0}^{\infty} i p_i \\
&= \sum_{i=0}^{\infty} i \rho^i(1-\rho) \\
&= \frac{\rho(1-\rho)}{(1-\rho)^2} = \frac{\rho}{1-\rho}.
\end{aligned}
$$

The expected number of customers waiting in the queue is given by

$$
\begin{aligned}
L_q &= \sum_{i=1}^{\infty} (i-1) p_i \\
&= \sum_{i=1}^{\infty} (i-1) \rho^i(1-\rho) \\
&= \frac{\rho}{1-\rho} - \rho.
\end{aligned}
$$

Moreover the expected number of customers in service is given by

$$L_s = 0 \cdot p_0 + 1 \cdot \sum_{i=1}^{\infty} p_i = 1 - (1-\rho) = \rho.$$

Clearly we have $L_c = L_q + L_s$. By using Little's queuing formula we have the average waiting time in the system, in the queue, and in service given respectively by

$$W_c = \frac{\rho}{\lambda(1-\rho)}, \quad W_q = \frac{\rho^2}{\lambda(1-\rho)}, \quad \text{and} \quad W_s = \frac{\rho}{\lambda}.$$

1.3.2 The n-Machine Interference Model

Machine interference occurs when less than one operator is allocated to operate one machine. Interested readers may consult [48, 54, 58, 65]. For example, one operator is in charge of two or more machines, or a group of r operators have to serve a set of m machines ($r < m$). In these situations, a machine may stand idle and wait for service (repair) when all the operators are busy. In this section we consider a system of n identical unreliable machines looked after by an operator. Broken machines are put in a queue waiting for repair by the operator. The mean normal time of a machine and the repair time of a machine are exponentially distributed with means being λ^{-1} and μ^{-1} respectively. The system is said to be in state $i(i = 0, 1, \ldots, n)$ if there are i broken machines. One can construct the following Markov chain for the model as follows (see Fig. 1.4). If we order the states of the system in increasing

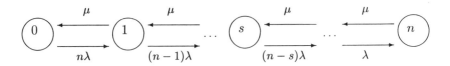

Fig. 1.4. The Markov chain for the n-machine interference model.

numbers of broken machines, it is not difficult to show that the generator matrix for this system is given by the following $(n+1) \times (n+1)$ tri-diagonal matrix $A_2 = A_{(n,\lambda,\mu)}$ where

$$A_2 = \begin{pmatrix} * & -\mu & & & 0 \\ -n\lambda & * & -\mu & & \\ & \ddots & \ddots & \ddots & \\ & & -2\lambda & * & -\mu \\ 0 & & & -\lambda & * \end{pmatrix}. \tag{1.41}$$

Here "*" is defined such that each column sum is zero. The underlying Markov chain is irreducible and the solution for the steady-state probability distribution can be shown to be

$$\mathbf{P}_{(n,\lambda,\mu)} = (p_0, p_1, \ldots, p_n)^t \tag{1.42}$$

where

$$p_i = \alpha \frac{n!}{(n-i)!} \left(\frac{\lambda}{\mu}\right)^i \quad \text{and} \quad \alpha^{-1} = \sum_{i=0}^{n} p_i. \tag{1.43}$$

Here p_i is the probability that there are i broken machines in the system.

Example 1.3.2. Using the steady-state probability distribution one can investigate the relationship between the number of normal machines in the system and the parameter $\rho = \lambda/\mu$. When the number of machines n is given, one may construct the following table.

Table 1.1. Expected number of normal machines

ρ	$n = 2$	$n = 4$	$n = 6$	$n = 10$
0.10	1.80	3.53	5.15	7.85
0.20	1.62	3.01	4.04	4.91
0.30	1.46	2.52	3.09	3.33
0.40	1.32	2.13	2.43	2.50

1.3.3 The Two-Queue Free System

Let us consider a more complicated queuing system. Suppose that there are two one-queue systems (as discussed in Section 1.3.1), see Fig. 1.5. The queuing system consists of two independent queues with the number of identical servers and waiting spaces being s_i and $n_i - s_i - 1$ $(i = 1, 2)$ respectively. Let the arrival rate of customers in the queue i be λ_i and service rate of the servers be μ_i $(i = 1, 2)$. Now the states of the queuing system can be represented by the elements in the set of ordered pair of integers

$$S = \{(i,j)|0 \leq i \leq n_1, 0 \leq j \leq n_2\}.$$

Here (i, j) represents the state that there are i customers in queue 1 and j customers in queue 2. This is a two-dimensional queuing model. If we order the states lexicographically, the generator matrix can be shown to be the following $n_1 n_2 \times n_1 n_2$ matrix in *tensor product* form [19, 20]:

$$A_3 = I_{n_1} \otimes A_{(n_2, s_2, \lambda_2, \mu_2)} + A_{(n_1, s_1, \lambda_1, \mu_1)} \otimes I_{n_2}. \tag{1.44}$$

Here $\otimes$ is the Kronecker tensor product; see Horn and Johnson [76]. The Kronecker tensor product of two matrices A and B of sizes $p \times q$ and $r \times s$ respectively is a $(pr) \times (qs)$ matrix given as follows:

$$A \otimes B = \begin{pmatrix} a_{11}B & \cdots & \cdots & a_{1q}B \\ a_{21}B & \cdots & \cdots & a_{2q}B \\ \vdots & \vdots & \vdots & \vdots \\ a_{p1}B & \cdots & \cdots & a_{pq}B \end{pmatrix}.$$

The Kronecker tensor product is a useful tool for representing generator matrices in our context.

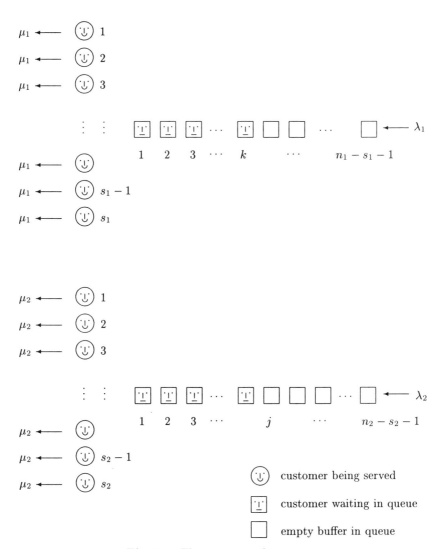

Fig. 1.5. The two-queue free system.

For this two-queue free queuing system, it is also not difficult to show that the steady-state probability distribution is given by the vector (see Exercise 1.9)

$$\mathbf{P}_{(n_1,s_1,\lambda_1,\mu_1)} \otimes \mathbf{P}_{(n_2,s_2,\lambda_2,\mu_2)}. \tag{1.45}$$

1.3.4 The Two-Queue Overflow System

Now let us add the following system dynamics to the two-queue free system discussed Section 1.3.3. We allow overflow of customers from queue 2 to queue 1 whenever queue 2 is full and there is still waiting space in queue 1; see Fig. 1.6. This is called the two-queue overflow system; see Kaufman [83] and Chan [19].

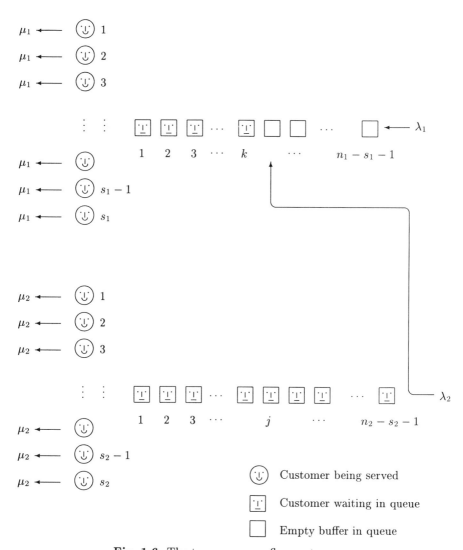

Fig. 1.6. The two-queue overflow system.

In this case, the generator matrix is given by the following matrix:

$$A_4 = I_{n_1} \otimes A_{(n_2,s_2,\lambda_2,\mu_2)} + A_{(n_1,s_1,\lambda_1,\mu_1)} \otimes I_{n_2} + R \otimes \mathbf{e_{n_2}}{}^t\mathbf{e_{n_2}}. \quad (1.46)$$

Here $\mathbf{e_{n_2}}$ is the unit vector $(0,0,\ldots,0,1)$ and

$$R = \begin{pmatrix} \lambda_2 & & & & 0 \\ -\lambda_2 & \lambda_2 & & & \\ & -\lambda_2 & \ddots & & \\ & & \ddots & \lambda_2 & \\ 0 & & & -\lambda_2 & 0 \end{pmatrix}. \quad (1.47)$$

In fact

$$A_4 = A_3 + R \otimes \mathbf{e_{n_2}}{}^t\mathbf{e_{n_2}},$$

where $R \otimes \mathbf{e_{n_2}}{}^t\mathbf{e_{n_2}}$ is the matrix describing the overflow of customers from queue 2 to queue 1. Unfortunately, there is no analytical solution for the generator matrix A_4. However, we can still get an approximated steady-state probability distribution when both n_1 and n_2 tend to infinity under some conditions.

Example 1.3.3. If $n_1 = n_2 = n$ and $\lambda_i/s_i\mu_i = \rho_i < 1$ then we have

$$A_4(\mathbf{p}_{(n_1,s_1,\lambda_1,\mu_1)} \otimes \mathbf{p}_{(n_2,s_2,\lambda_2,\mu_2)}) = R \otimes \mathbf{e_{n_2}}{}^t\mathbf{e_{n_2}}(\mathbf{p}_{(n_1,s_1,\lambda_1,\mu_1)} \otimes \mathbf{p}_{(n_2,s_2,\lambda_2,\mu_2)})$$
$$= (R\mathbf{e_{n_2}}(\mathbf{p}_{(n_1,s_1,\lambda_1,\mu_1)})) \otimes (\mathbf{e_{n_2}}{}^t\mathbf{e_{n_2}}\mathbf{p}_{(n_2,s_2,\lambda_2,\mu_2)}).$$

We note that

$$||R\mathbf{e_{n_2}}\mathbf{p}_{(n_1,s_1,\lambda_1,\mu_1)}||_\infty \le \lambda_2$$

and by (1.40)

$$||\mathbf{e_{n_2}}{}^t\mathbf{e_{n_2}}\mathbf{p}_{(n_2,s_2,\lambda_2,\mu_2)}||_\infty = \alpha^{-1} \prod_{k=1}^{n_2} \frac{\lambda}{\mu\min\{k,s_2\}}.$$

Here

$$||\mathbf{y}||_\infty = ||(y_1,y_2,\ldots,y_n)^t||_\infty = \max_i\{|y_i|\}.$$

Since

$$\alpha^{-1} = \sum_{i=0}^{s_2-1}\left(\prod_{k=1}^{i+1}\frac{\lambda}{\mu\min\{k,s_2\}}\right) + \left(\frac{\lambda}{\mu}\right)^{s+1}\frac{1}{s!}(\rho_2 + \rho_2^2 + \ldots + \rho_2^{n_2-s_2-1}),$$

we have

$$\lim_{n_2\to\infty}\alpha^{-1} = \sum_{i=0}^{s_2-1}\left(\prod_{k=1}^{i+1}\frac{\lambda}{\mu\min\{k,s_2\}}\right) + \left(\frac{\lambda}{\mu}\right)^{s+1}\frac{1}{s!}\frac{\rho_2}{1-\rho_2} < \infty.$$

Moreover we have

$$\lim_{n_2 \to \infty} \prod_{k=1}^{n_2} \frac{\lambda}{\mu \min\{k, s_2\}} = 0,$$

and therefore

$$\lim_{n_2 \to \infty} ||\mathbf{e_{n_2}}^t \mathbf{e_{n_2}} \mathbf{P_{(n_2,s_2,\lambda_2,\mu_2)}}||_\infty = 0.$$

We remark that the above limit tends to zero at a rate of $\rho_2^{n_2}$ as n_2 goes to infinity. Hence we have

$$\lim_{n_2 \to \infty} ||A_4(\mathbf{P_{(n,s_1,\lambda_1,\mu_1)}} \otimes \mathbf{P_{(n,s_2,\lambda_2,\mu_2)}})||_\infty \le \lambda_2 \cdot 0 = 0.$$

This means that

$$\mathbf{P_{(n_1,s_1,\lambda_1,\mu_1)}} \otimes \mathbf{P_{(n_2,s_2,\lambda_2,\mu_2)}}$$

is an approximated steady-state probability vector for the two-queue overflow model for large values of n_1 and n_2.

1.4 Iterative Methods

Usually the steady-state probability distributions of queuing and manufacturing systems do not have analytical solutions and their generator matrices are large and sparse. Therefore the cost of direct methods for solving the steady-state probability distribution is too large when the problem size increases. Very often they are solved by *iterative methods*; see [9, 10, 51, 101]. There are quite a number of common iterative methods (Gauss-Seidel method, Jacobi method, etc.) for solving linear systems, see for instance Axelsson [5], Golub and van Loan [70], and Varga [113]. Here in our discussion we will concentrate on the Conjugate Gradient (CG) method. By using the technique of preconditioning, the convergence rate of the CG method can be far superior to classical iterative methods, see for instance [37, 38, 59, 60, 89].

To apply an iterative method in solving the steady-state probability vector of an irreducible generator matrix A, one should aware that the matrix A is singular and has a null space of dimension one. For the sake of presentation, we assume that A has positive diagonal entries, non-positive off-diagonal entries, and each column sum of A is zero. Thus

$$(I - A \cdot (\text{diag}(A))^{-1})$$

is a *stochastic matrix*. By using the fact that the generator matrix A is irreducible and the Perron and Frobenius theory [113], the generator matrix A has a unique steady-state probability vector $\mathbf{p}$. We are interested in solving the linear system

$$A\mathbf{p} = \mathbf{0} \quad \text{and} \quad \sum_i p_i = 1.$$

There are two methods to solve for the vector $\mathbf{p}$. The first method is to consider the following linear system which has unique solution

$$G_1\mathbf{x} = (A + \mathbf{e_1}{}^t\mathbf{e_1})\mathbf{x} = \mathbf{e_1} \tag{1.48}$$

where $\mathbf{e}_1 = (1, 0, \ldots, 0)$. The matrix G is non-singular because it is irreducible diagonally dominant with the last column being strictly diagonal dominant. It can be shown that G_1 is an invertible matrix and the solution takes the following form:

$$\mathbf{x} = (x_1, x_2, \ldots, x_{n-1}, 1)^t$$

and

$$(A + \mathbf{e_1}{}^t\mathbf{e_1})\mathbf{x} = A\mathbf{x} + \mathbf{e_1} = \mathbf{e_1}.$$

Thus the required probability distribution vector is given by

$$\mathbf{p} = (\mathbf{1} \cdot \mathbf{x})^{-1}\mathbf{x} \quad \text{where} \quad \mathbf{1} = (1, 1, \ldots, 1).$$

The second method is to consider a reduced linear system as follows. Let

$$A_{n-1} = \begin{pmatrix} a_{1,1} & \cdots & \cdots & a_{1,n-1} \\ a_{2,1} & \cdots & \cdots & a_{2,n-1} \\ \vdots & \vdots & \vdots & \vdots \\ a_{n-1,1} & \cdots & \cdots & a_{n-1,n-1} \end{pmatrix} \tag{1.49}$$

be the sub-matrix of A obtained by deleting the last row and last column of A and

$$\mathbf{b} = -(a_{1,n}, a_{2,n}, \ldots, a_{n-1,n})^t.$$

By solving the linear system

$$G_2\mathbf{x} = A_{n-1}\mathbf{x} = \mathbf{b} \tag{1.50}$$

the steady-state probability vector can be obtained as follows:

$$\mathbf{p} = (\mathbf{1} \cdot \mathbf{x} + 1)^{-1}(\mathbf{x}, 1) \quad \text{where} \quad \mathbf{1} = (1, 1, \ldots, 1).$$

We note that the first method preserves the structure of the generator matrix A, while the second method reduces the number of linear equations by one.

1.4.1 Classical Iterative Methods

Classical iterative methods such as the *Jacobi method* and the *Gauss Seidel method* can be derived quite easily. Suppose we are going to solve the following linear system $A\mathbf{x} = \mathbf{b}$ where

$$A = \begin{pmatrix} a_{1,1} & \cdots & \cdots & a_{1,n} \\ a_{2,1} & \cdots & \cdots & a_{2,n} \\ \vdots & \vdots & \vdots & \vdots \\ a_{n,1} & \cdots & \cdots & a_{n,n} \end{pmatrix}, \tag{1.51}$$

$$\mathbf{x} = (x_1, x_2, \ldots, x_n)^t,$$

and

$$\mathbf{b} = (b_1, b_2, \ldots, b_n)^t.$$

The matrix A can be split in the following way:

$$A = L + D + U$$

where

$$L = \begin{pmatrix} 0 & 0 & \cdots & & \cdots & 0 \\ a_{2,1} & 0 & \ddots & & \ddots & \vdots \\ \vdots & \ddots & \ddots & & 0 & 0 \\ a_{n,1} & \cdots & & \cdots & a_{n,n-1} & 0 \end{pmatrix}, \tag{1.52}$$

$$D = \mathrm{Diag}(a_{1,1}, a_{2,2}, \cdots, a_{n,n}),$$

and

$$U = \begin{pmatrix} 0 & a_{1,2} & \cdots & & \cdots & a_{1,n} \\ 0 & 0 & a_{2,3} & & \ddots & \vdots \\ \vdots & \ddots & \ddots & & \ddots & a_{n-1,n} \\ 0 & \cdots & & \cdots & 0 & 0 \end{pmatrix}. \tag{1.53}$$

Let $\mathbf{x}_0$ be an initial guess solution of $A\mathbf{x} = \mathbf{b}$ and $\mathbf{x}_k$ be the approximated solution obtained in the kth iteration. Then the Jacobi method reads (assume D^{-1} exists)

$$\mathbf{x}_{k+1} = D^{-1}(\mathbf{b} - (L+U)\mathbf{x}_k) \quad \text{for } k = 0, 1, \ldots \tag{1.54}$$

and the Gauss-Seidel method reads (assume $(L+D)^{-1}$ exists)

$$\mathbf{x}_{k+1} = (L+D)^{-1}(\mathbf{b} - U\mathbf{x}_k) \quad \text{for } k = 0, 1, \ldots. \tag{1.55}$$

Let $J = D^{-1}(L+U)$ and $S = (D+L)^{-1}U$; they are the *iteration matrices* for the Jacobi method and Gauss-Seidel method respectively. Then (1.54) and (1.55) can be written respectively as

$$\mathbf{x}_{k+1} = (I - J + J^2 + \cdots + (-J)^k)D^{-1}\mathbf{b} + (-J)^{k+1}\mathbf{x}_0 \tag{1.56}$$

and

$$\mathbf{x}_{k+1} = (I - S + S^2 + \cdots + (-S)^k)(D+L)^{-1}\mathbf{b} + (-S)^{k+1}\mathbf{x}_0. \tag{1.57}$$

We note that if $\rho(J) < 1$ then we have

$$\lim_{k \to \infty} (-J)^k = 0.$$

Here
$$\rho(J) = \max\{|\lambda| : \lambda \in \lambda(J)\}$$
is called the *spectral radius* and $\lambda(J)$ is the set containing all the eigenvalues of J. Thus if $\rho(J) < 1$, then we have

$$\lim_{k\to\infty} (I - J + J^2 + \cdots + (-J)^k)(I + J) = \lim_{k\to\infty} I + (-J)^k J = I.$$

This implies that

$$\lim_{k\to\infty} (I - J + J^2 + \cdots + (-J)^k) = (I + J)^{-1}.$$

This means that the Jacobi method converges to the solution of $A\mathbf{x} = \mathbf{b}$. Similarly if $\rho(S) < 1$ then we have

$$\lim_{k\to\infty} (I - S + S^2 + \cdots + (-S)^k) = (I + S)^{-1}$$

and the Gauss-Seidel method converges.

We remark that some iterative methods can be constructed by splitting the matrix A into two parts. There are a lot of methods for splitting $A = M + N$ into two parts M and N; see for instance [5, 70, 113]. For the Jacobi method $M = D$ and $N = L + U$ and for the the Gauss-Seidel method $M = L + D$ and $N = U$. But a splitting should satisfy the following two conditions:

(B1) $M\mathbf{y} = \mathbf{r}$ can be solved easily, and
(B2) $\rho(M^{-1}N) < 1$.

When the Jacobi method or Gauss-Seidel method is applied to solving our generator matrix system (1.48) or (1.49), the spectral radius is not necessarily smaller than one. However, if the underlying Markov chain is irreducible then the spectral radius of the iteration matrix is smaller than one (see Horn and Johnson [76, p. 363]). Hence both the Jacobi method and Gauss-Seidel method converge.

1.4.2 Conjugate Gradient Method and Its Convergence Rate

In this book we focus on Conjugate Gradient (CG) type methods for solving the steady-state probability distributions of the captured queuing and manufacturing systems. Given a Hermitian, positive definite $n \times n$ matrix H_n, a well-known and successful iterative method for solving the linear system $H_n\mathbf{x} = \mathbf{b}$ is the CG method. This method was first discussed by Hestenes and Stiefel [73] as a direct method (see also Golub and van Loan [70] and Axelsson and Barker [6]). It has become a popular iterative solver for linear systems over the past two decades and has superseded the classical iterative methods such as the Jacobi and Gauss-Seidel. The CG method reads:

Choose an initial guess $\mathbf{y}_0$;
$\mathbf{r}_0 = \mathbf{b} - H\mathbf{y}_0$;
$k = 1$;
$\mathbf{p}_1 = \mathbf{r}_0$;
$\alpha_1 = \mathbf{r}_0^t \mathbf{r}_0 / \mathbf{p}_1^t H \mathbf{p}_1$;
$\mathbf{y}_1 = \mathbf{y}_0 + \alpha_1 \mathbf{p}_1$;
$\mathbf{r}_1 = \mathbf{r}_0 - \alpha_1 H \mathbf{p}_1$;
while $\|\mathbf{r}_k\|_2 > tolerance$,
$k = k + 1$;
$\beta_k = \mathbf{r}_{k-1}^t \mathbf{r}_{k-1} / \mathbf{r}_{k-2}^t \mathbf{r}_{k-2}$;
$\mathbf{p}_k = \mathbf{r}_{k-1} + \beta_k \mathbf{p}_{k-1}$;
$\alpha_k = \mathbf{r}_{k-1}^t \mathbf{r}_{k-1} / \mathbf{p}_k^t H \mathbf{p}_k$;
$\mathbf{y}_k = \mathbf{y}_{k-1} + \alpha_k \mathbf{p}_k$;
$\mathbf{r}_k = \mathbf{r}_{k-1} - \alpha_k H \mathbf{p}_k$;
end;
$\mathbf{y} = \mathbf{y}_k$.

The convergence rate of this method depends on the spectrum of the matrix H_n; see Axelsson and Barker [6]. For example if the spectrum of H_n is contained in an interval, i.e. $\sigma(H_n) \subseteq [a, b]$ and $\mathbf{x}_i$ is the approximated solution obtained in the ith iteration, then the error $\mathbf{r}_i = \mathbf{b} - H_n \mathbf{x}_i$ is given by

$$\frac{\|\mathbf{r}_i\|_2}{\|\mathbf{r}_0\|_2} \le 2 \left(\frac{\sqrt{b} - \sqrt{a}}{\sqrt{b} + \sqrt{a}} \right)^i \tag{1.58}$$

i.e. the convergence rate is linear. Hence the approximate upper bound for the number of iterations required to make the relative error

$$\frac{\|\mathbf{r}_i\|_2}{\|\mathbf{r}_0\|_2} \le \delta \tag{1.59}$$

is given by

$$\frac{1}{2} \left(\sqrt{\frac{b}{a}} - 1 \right) \log \left(\frac{2}{\delta} \right) + 1. \tag{1.60}$$

When the convergence rate of the method is not good, then it can be improved by using a *preconditioner*. A good preconditioner C is a matrix satisfying the following criteria [5, 6]:

(C1) C can be constructed easily;
(C2) for a given right hand side vector $\mathbf{r}$, the linear system $C\mathbf{y} = \mathbf{r}$ can be solved easily; and
(C3) the spectrum (or singular values) of the preconditioned system $C^{-1} H_n$ should be clustered around one.

In fact, in the coming chapters we will apply condition (C3) in proving convergence rate. In the following we give the definition of clustering.

Definition 1.4.1. *We say that a sequence of matrices S_n of size n has a clustered spectrum around one if for all $\epsilon > 0$, there exist non-negative integers n_0 and n_1, such that for all $n > n_0$, at most n_1 eigenvalues of the matrix $S_n^* S_n - I_n$ have absolute values larger than ϵ.*

One sufficient condition for the matrix to have eigenvalues clustered around one is that

$$H_n = I_n + L_n,$$

where I_n is the $n \times n$ identity matrix and L_n is a low rank matrix ($\text{Rank}(L_n)$ is bounded above and independent of the matrix size n).

1.4.3 The Conjugate Gradient Squared Method

We note that in general the generator matrices of our queuing and manufacturing systems are neither Hermitian nor positive definite. To handle this problem, one may consider the normal equation of the original linear system,

$$G^t G \mathbf{x} = G^t \mathbf{b}.$$

This approach is called the CGNR (Conjugate Gradient on the Normal equations to minimize the Residual). The advantage of this approach is that all the theory for CG method can be applied to the normal equation. But one important disadvantage is that it involves the vector multiplication of the conjugate transpose of the original matrix.

There are a number of transpose free CG methods for solving non-symmetric linear systems; see Axelsson [5] and Kelley [84] for instance. Here we consider a generalized conjugate gradient method, namely the Conjugate Gradient Squared (CGS) method; see Sonneveld [107]. Let $\mathbf{x_0}$ be the initial guess of the solution of the non-singular system $B\mathbf{x} = \mathbf{b}$, then the CGS algorithm reads:

$\mathbf{x} = \mathbf{x_0}$;
$\mathbf{r}_0 = \mathbf{b} - B\mathbf{x}$;
$\tilde{\mathbf{r}} = \mathbf{r}_0$;
$\mathbf{s} = \mathbf{r}_0$;
$\mathbf{p} = \mathbf{r}_0$;
$\mathbf{r} = \mathbf{r}_0$;
$\mathbf{w} = B\mathbf{p}$;
$\mu = \tilde{\mathbf{r}}^t \mathbf{r}$;
repeat the following :
$\quad \gamma = \mu$;
$\quad \alpha = \gamma / \tilde{\mathbf{r}}^t \mathbf{r}$;

$\mathbf{q} = \mathbf{s} - \alpha\mathbf{w};$
$\mathbf{d} = \mathbf{s} + \mathbf{q};$
$\mathbf{w} = B\mathbf{d};$
$\mathbf{x} = \mathbf{x} + \alpha\mathbf{d};$
$\mathbf{r} = \mathbf{r} - \alpha\mathbf{w};$
if converge ($||\mathbf{r}||_2/||\mathbf{r}_0||_2 <$ tolerance) then exit
$\mu = \tilde{\mathbf{r}}^t\mathbf{r};$
$\beta = \mu/\gamma;$
$\mathbf{s} = \mathbf{r} - \beta\mathbf{q};$
$\mathbf{p} = \mathbf{s} + \beta(\mathbf{q} + \beta\mathbf{p});$
$\mathbf{w} = B\mathbf{p};$
end;

We see that the algorithm does not involve the vector multiplication of the form $B^t\mathbf{y}$. In each iteration, it requires only two inner products and matrix multiplication of the form $B\mathbf{y}$. In the Preconditioned Conjugate Gradient Squared (PCGS) method with preconditioner C, we applied the CGS algorithm to solve the preconditioned linear system $C^{-1}B\mathbf{x} = C^{-1}\mathbf{b}$.

Here we would like to mention that one of the earliest applications of PCG methods in solving queuing networks was done by Chan [19, 20]. For Markovian overflow networks with traffic density close to one, the generator matrices are close to the discretization matrices of elliptic equations. Using techniques from elliptic equations, such as the fast Poisson solvers and domain decomposition methods [23], Chan has constructed efficient preconditioners for these networks. These preconditioners make use of the tensor structure of the generator matrices and are easy to construct and invert.

Example 1.4.1. Let us consider a Markovian one-machine repairing model as follows. The system consists of an unreliable machine; the normal time of the machine is exponentially distributed with parameter λ. Once the machine is broken, it is subject to an n-phase repairing process. The repairing time at phase i is also exponentially distributed with parameter $\mu_i(i = 1, 2, \ldots, n)$. We let 0 be the state where the machine is normal and i be the state where the machine is in repairing phase i. The Markov chain of the model is given by Fig. 1.7. According to the ordering of the states, the generator matrix for

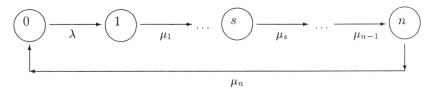

Fig. 1.7. The Markov chain for the one-machine repairing system.

the system is given by the following $(n + 1) \times (n + 1)$ matrix:

$$A_5 = \begin{pmatrix} \lambda & 0 & & & & -\mu_n \\ -\lambda & \mu_1 & 0 & & & \\ & -\mu_1 & \ddots & \ddots & & \\ & & & \ddots & \mu_{n-1} & 0 \\ 0 & & & & -\mu_{n-1} & \mu_n \end{pmatrix}. \tag{1.61}$$

We test the performance of the iterative methods in solving the steady-state probability vector. For simplicity, we let $\lambda = \mu = 1$ in the following numerical examples. Thus in this case, the generator matrix takes the following form:

$$A_6 = \begin{pmatrix} 1 & 0 & & & & -1 \\ -1 & 1 & 0 & & & \\ & -1 & \ddots & \ddots & & \\ & & & \ddots & 1 & 0 \\ 0 & & & & -1 & 1 \end{pmatrix}. \tag{1.62}$$

It is obvious that the steady-state probability vector for A_6 is given by

$$\mathbf{p} = \frac{1}{n+1}(1, 1, \ldots, 1).$$

One possible preconditioner for A_6 is the following bidiagonal matrix:

$$C = \begin{pmatrix} 1 & 0 & & & & 0 \\ -1 & 1 & 0 & & & \\ & -1 & \ddots & \ddots & & \\ & & & \ddots & 1 & 0 \\ 0 & & & & -1 & 1 \end{pmatrix} \tag{1.63}$$

which is the lower triangular part of the matrix A_6. There is no construction cost for C and clearly $C\mathbf{x} = \mathbf{r}$ can be solved easily in $O(n)$ operations. Moreover, Rank$(C - A_6) = 1$, therefore C is a good approximation to A_6.

We are going to apply the Jacobi (J) method, the Gauss-Seidel (GS) method, the CGS method, and the PCGS method with preconditioner C to solving the steady-state probability vector of A_6. The stopping criterion is

$$\|\mathbf{b} - G_1\mathbf{x}\|_2 < 10^{-7}$$

for all the methods for the sake of comparison. The convergence rate of the CGS, and PCGS methods does not depend much on the initial guess. To demonstate this we use a random vector generated by MATLAB (with seed = 1) as the initial guess. Table 1.2 and Table 1.3 show the numerical results. The symbols N_{CGS} and N_{PCGS} represent the number of iterations for convergence by the CGS and the PCGS with preconditioner C respectively.

The cost per iteration for all methods is of the same order $O(n)$. We note that PCGS converges in two steps, while the GS method converges in around 23 iterations. The convergence rate of both the CGS method and Jacobi method increase like $O(n)$. Therefore the total computational costs for CGS method and Jacobi method are of $O(n^2)$, while the costs for both the PCGS method, and GS method are $O(n)$. Moreover we note that both the PCGS method and the GS method give a better solution $\mathbf{x}_{N_{\mathrm{PCGS}}}$ in the sense that the error $||\mathbf{x}_{N_{\mathrm{PCGS}}} - \mathbf{p}||_\infty$ is of smaller error. We remark that the iteration matrix S of the GS method is a rank one matrix and $\rho(S) = 0.5$ (see Exercise 1.10). This explains why the GS method converges so fast.

Table 1.2. Number of iterations for convergence

n	Matrix size	N_{CGS}	$||\mathbf{x}_{N_{\mathrm{CGS}}} - \mathbf{p}||_\infty$	N_{PCGS}	$||\mathbf{x}_{N_{\mathrm{PCGS}}} - \mathbf{p}||_\infty$
9	10	10	10^{-13}	2	10^{-17}
19	20	19	10^{-12}	2	10^{-17}
39	40	39	10^{-10}	2	10^{-18}
79	80	79	10^{-11}	2	10^{-18}
159	160	164	10^{-11}	2	10^{-18}
319	320	331	10^{-11}	2	10^{-18}

Table 1.3. Number of iterations for convergence

n	Matrix size	N_J	$||\mathbf{x}_{N_J} - \mathbf{p}||_\infty$	N_{GS}	$||\mathbf{x}_{N_{\mathrm{GS}}} - \mathbf{p}||_\infty$
9	10	263	10^{-9}	22	10^{-17}
19	20	498	10^{-9}	22	10^{-17}
39	40	1019	10^{-9}	22	10^{-18}
79	80	2021	10^{-10}	23	10^{-17}
159	160	4151	10^{-10}	24	10^{-17}
319	320	8410	10^{-11}	24	10^{-17}

1.5 Outline of the Book

In Chapter 2, a queuing system with batch arrival of demand is presented. The Preconditioned Conjugate Gradient (PCG) method is applied to solve for the steady-state probability distribution vector. Toeplitz-circulant pre-conditioner is constructed by exploiting the near Toeplitz structure of the system generator matrix. We compare our method with some classical iterative methods. We show that the preconditioned matrix has singular values clustered around one. We expect that the proposed method converges very fast when the conjugate gradient method is applied to solving the preconditioned system. Numerical results are given to illustrate the fast convergence.

In Chapter 3, we consider a queuing system with Markov-modulated Poisson process (MMPP) inputs. The MMPP is a generalization of Poisson processes and is commonly used in modeling the input process of communication systems such as data traffic systems and ATM networks. The generator matrices of these processes are tri-diagonal block matrices with each diagonal block being a sum of tensor products of matrices. We are interested in finding the steady-state probability distributions of these processes which are the normalized null-vectors of their generator matrices. A classical iterative method, such as the Block Gauss-Seidel (BGS) method is usually employed to solve for the steady-state probability distributions. Classical iterative methods are easy to implement, but their convergence rates are slow in our experience. The number of iterations required for convergence increases like $O(n)$ where n is the size of the waiting spaces in the queues. The PCG method is applied to solve the problem. We construct our preconditioners by taking circulant approximations of the tensor blocks of the generator matrices. Numerical results are given to illustrate the fast convergence.

As an application of the results in Chapter 3, in Chapter 4 we apply the MMPP to model manufacturing systems of multiple unreliable machines. The production process consists of multiple parallel machines which produce one type of product. Each machine has exponentially distributed normal time, down time and processing time for one unit of product. The inter-arrival of a demand is exponentially distributed and finite backlog is allowed. We employ Hedging Point Production (HPP) policy as the production control. The average running cost of the system can be written in terms of the steady-state probability distribution. Our iterative algorithm developed for the queuing systems in Chapter 3 can be applied to obtain the steady-state probability distribution for the system and hence the optimal hedging point.

In Chapter 5, we study optimal HPP policy for a failure prone one-machine manufacturing system. The machine produces one type of product and its demand has batch arrival. The inter-arrival time of the demand, normal time of the machine, and processing time for one unit of product are exponentially distributed. When the machine is down, it is subject to a sequence of l repairing phases (the general Erlangian distribution). In each phase, the repairing time is exponentially distributed. The machine states and the inventory levels are modeled as Markovian processes. The PCG method is presented to solve the steady-state probability distribution. The average running cost for the system can be written in terms of the steady-state probability distribution. The optimal hedging point can then be obtained by varying different values of the hedging point.

In Chapter 6, a Markovian model of Flexible Manufacturing Systems (FMS) is presented. We consider FMS of multiple unreliable machines producing one type of product. When a machine breaks down, it is subject to a repairing process if there is a maintenance facility available, otherwise it will queue up and wait for repair. HPP policy is employed as the production control. A circulant-based preconditioner is constructed to solve the steady-state probability distribution for the FMS. Then the system performance such as the average system running cost, system throughput, and machine utilization rate can be written in terms of the probability distribution. Moreover, the average profit for the system can also be written in terms of the probability distribution. Therefore the optimal hedging point can be obtained by varying different possible values of hedging points.

In Chapter 7, we study a two-stage manufacturing system under HPP policy. The manufacturing system produces one type of product and its demand is modeled as a Poisson process. Each product has to go through the manufacturing process by a reliable machine in each stage. The PCG method is employed to solve the steady-state probability distribution of the system. A preconditioner is constructed by taking a circulant approximation of the generator matrix of the system. Numerical examples are given to demonstrate the fast convergence rate of the proposed method. Extension to multiple parallel machines in each stage and numerical approximation for the steady-state probability distribution are also discussed.

The delivery time guarantee policy is used in commercial companies as a marketing strategy to attract customers. For example, United Parcel Services guarantees the next day delivery by 8:30 am. Lucky, a major supermarket chain in California, use a "three's a crowd" campaign, which guarantees a new checkout counter will be opened if there are more than three people waiting in a checkout queue. Wells Fargo Bank offers a "five minute maximum wait policy" which offers five dollars to the customer if the customer waits for more than five minutes in line. In Chapter 8, we discuss an analytical model for a manufacturing system under delivery time guarantee policy. An analytical solution of the steady-state probability distribution is obtained and is applied to the analysis of the model.

In Chapter 9, two inventory models are discussed. In the first model, we apply the MMPP to model inventory systems of multi-location. It is common to allow Emergency Lateral Transshipments (ELTs) from local locations to the main depot. Here we propose a new model for the inventory system of consumable items. The inventory system of each location and the main depot is modeled by Markovian queuing networks. The transshipments are modeled by MMPP processes. An asymptotic solution of the system steady-state probability distribution is derived for the case where the number of transshipments is large. We then consider a two-echelon inventory model. The inventory system consists of an infinite supply plant, a central warehouse, and a number of local warehouses. Since the customers are usually scattered over a large regional area, a network of inventory locations (local warehouses) is necessary to maintain a high service level. In particular, in our study ELT is allowed among the local warehouses to enhance the service level.

Exercises

1.1 Suppose that calls arriving to a station follow a Poisson process with mean rate λ. Since the station is out of order, the calls will then congregate there. At the beginning there are k calls at the station. What is the probability that there are $m(m \geq k)$ calls aggregated at the station after a time t?

1.2 Show that

$$\begin{pmatrix} 1-q & r \\ q & 1-r \end{pmatrix} = \begin{pmatrix} r & 1 \\ q & -1 \end{pmatrix} \begin{pmatrix} 1 & 0 \\ 0 & 1-q-r \end{pmatrix} \begin{pmatrix} r & 1 \\ q & -1 \end{pmatrix}^{-1}.$$

Hence show that (1.5) is the solution for (1.4).

1.3 Using the method of mathematical induction to prove that (1.24) is the solution for (1.19).

1.4 Let

$$F(x) = \int_x^\infty f(t)dt$$

and $f(t)$ be a continuous probability density function such that $f(x) = 0$ for $x < 0$. Suppose that

$$F(x+h) = F(x)F(h)$$

for any non-negative x and h, show that $f(t)$ is the exponential distribution. Hint: note that

$$F'(x) = \lim_{h \to 0} \frac{F(x+h) - F(x)}{h} = F(x) \left(\lim_{h \to 0} \frac{F(h) - F(0)}{h} \right).$$

1.5 Beginning from the first two rows of A_1 in (1.38), perform the following row operations until the second to last row of A_1: add the ith row to the $(i+1)$th row. Show that (1.39) is the unique normalized null vector of A_1.

1.6 Find the expected number of customers in the queuing system discussed in Section 1.3.1 when the number of servers is one. Apply Little's queuing formula to find the average waiting time of customers in system.

1.7 Show that the steady-state probability distribution of the generator matrix in (1.41) is given by (1.42).

1.8 Consider the machine interference model discussed in Section 1.3.2; what is the steady-state probability that the operator is idle when $n = 6$ and $\lambda/\mu = 0.5$?

1.9 Use the formula $(A \otimes B)(P \otimes Q) = (AP) \otimes (BQ)$ to show that the vector (1.45) is the steady-state probability distribution vector for (1.44).

1.10 Show that in the machine repairing problem in Example 1.4.1, the iteration matrix S of the Gauss-Seidel method is a rank one matrix and $\rho(S) = 0.5$.

1.11 Prove or disprove the following:
(a) $\rho(AB) \leq \rho(A)\rho(B)$;
(b) $\rho(C + D) \leq \rho(C) + \rho(D)$.
Hint: consider

$$A = \begin{pmatrix} 1 & 1 \\ 0 & 1 \end{pmatrix}, \quad B = \begin{pmatrix} 1 & 0 \\ 1 & 1 \end{pmatrix}$$

and

$$C = \begin{pmatrix} 0 & 1 \\ 0 & 0 \end{pmatrix}, \quad D = \begin{pmatrix} 0 & 0 \\ 1 & 0 \end{pmatrix}.$$

2. Toeplitz-Circulant Preconditioners for Queuing Systems with Batch Arrivals

2.1 Introduction

In this chapter, we discuss Markovian queuing systems with batch arrivals. The system is similar to the one-queue system discussed in Chapter 1 except that the arrival process is not a simple Poisson process. This kind of queuing system occurs in many applications, for example telecommunication networks [94] and the loading dock models [102]. In the queuing system, we assume that the arrival process is an exogenous Poisson batch arrival process with mean batch inter-arrival time being λ^{-1}. For $i \geq 1$, we let

$$\lambda_i = q_i \lambda$$

where q_i is the probability that the batch size is i. We note that λ_i is the batch arrival rate for batches with size i and that

$$\lambda = \sum_{k=1}^{\infty} \lambda_k. \qquad (2.1)$$

The queuing system has s servers and each of them has exponential service time of mean μ^{-1}. The waiting space is of size $(n - s - 1)$. This means that if the arrival batch size is larger than the waiting place left, then only part of the arrival batch will be accepted, and the other customers will be cleared from the system.

We let the number of customers be the state of the system, then by ordering the state space lexicographically, the queuing system can be characterized by the following generator matrix A_n:

$$
A_n = \begin{pmatrix}
\lambda & -\mu & 0 & 0 & 0 & \cdots & 0 \\
-\lambda_1 & \lambda + \mu & -2\mu & 0 & 0 & \cdots & 0 \\
-\lambda_2 & -\lambda_1 & \lambda + 2\mu & \ddots & \ddots & & \vdots \\
\vdots & -\lambda_2 & \ddots & \ddots & -s\mu & \ddots & \\
& \vdots & \ddots & \ddots & \lambda + s\mu & \ddots & 0 \\
-\lambda_{n-2} & -\lambda_{n-3} & \cdots & & \ddots & \ddots & -s\mu \\
-r_1 & -r_2 & -r_3 & \cdots & -r_{s+1} & \cdots & s\mu
\end{pmatrix}, \qquad (2.2)
$$

where r_i are such that each column sum of A_n is zero, i.e.

$$r_i = \lambda - \sum_{k=n-i}^{\infty} \lambda_k.$$

Clearly the matrix A_n has zero column sum, positive diagonal entries, and non-positive off-diagonal entries. Thus if $\text{diag}(A_n)$ denotes the diagonal matrix containing the diagonal part of A_n, then the matrix

$$(I - A_n \cdot (\text{diag}(A_n))^{-1})$$

is a stochastic matrix. We now claim that the matrix A_n is irreducible. If $\lambda_i = 0$ for all $i = 1, \ldots, n-2$, then $r_1 = \lambda$ and the matrix is irreducible. Now if they are not all zero, say λ_j is the first non-zero λ_i, then $r_{n-j} = \lambda$. Hence A_n is also irreducible. From the Perron and Frobenius theory [113], A_n has a one-dimensional null-space with a positive null vector.

Since A_n has a one-dimensional null space, the steady-state probability vector $\mathbf{p}$ can be found by deleting the last column and the last row of A_n and solving the $(n-1) \times (n-1)$ reduced linear system

$$Q_{n-1}\mathbf{y} = (0, 0, \ldots, 0, s\mu)^t. \tag{2.3}$$

After getting $\mathbf{y}$, $\mathbf{p}$ is obtained by normalizing the vector $(\mathbf{y}, 1)^t$. This is exactly the second method discussed in Section 1.5. We choose this method because we can obtain a near-Toeplitz structure matrix Q_{n-1}. We will see how this can help us in the construction of preconditioners shortly.

Thus let us concentrate on solving non-homogeneous systems of the form

$$Q_n\mathbf{x} = \mathbf{b} \tag{2.4}$$

where

$$Q_n = \begin{pmatrix} \lambda & -\mu & 0 & 0 & 0 & \cdots & 0 \\ -\lambda_1 & \lambda+\mu & -2\mu & 0 & 0 & \cdots & 0 \\ -\lambda_2 & -\lambda_1 & \lambda+2\mu & \ddots & \ddots & & \vdots \\ \vdots & -\lambda_2 & \ddots & \ddots & -s\mu & \ddots & \vdots \\ \vdots & \vdots & \ddots & \ddots & \lambda+s\mu & \ddots & 0 \\ -\lambda_{n-2} & -\lambda_{n-3} & & \ddots & \ddots & \ddots & -s\mu \\ -\lambda_{n-1} & -\lambda_{n-2} & \cdots & \cdots & -\lambda_2 & -\lambda_1 & \lambda+s\mu \end{pmatrix}. \tag{2.5}$$

We note that if all of the λ_i, $i = 1, \ldots, n-1$ are zeros, then Q_n will be a bidiagonal matrix and can easily be inverted. Therefore in the following, we assume that at least one of the λ_i is non-zero. Clearly, Q_n is an irreducibly diagonally dominant matrix. In particular, if the system (2.4) is solved by classical iterative methods such as the Jacobi or the Gauss-Seidel methods, both methods will be convergent for all initial guesses.

We will see in Section 2.2.1 that the costs per iteration of the Jacobi and the Gauss-Seidel methods are $O(n \log n)$ and $O(n^2)$ respectively. The memory requirement is $O(n)$ for both methods. We remark that the system (2.4) can also be solved by Gaussian elimination in $O(n^2)$ operations with $O(n^2)$ memory. In the following, we are interested in solving (2.4) by conjugate gradient (CG) type methods. We will see that the cost per iteration of the method is $O(n \log n)$ and memory requirement is $O(n)$, the same as the Jacobi method. However, we are able to show that for sufficiently large n, the CG type method converges very fast and the total cost of solving (2.4) is around $O(n \log n)$ operations.

It has been mentioned in Chapter 1 that CG methods can be used to solve Hermitian positive definite system $H\mathbf{y} = \mathbf{r}$. The convergence rate of the method depends on the spectrum of the matrix H. The more clustered the eigenvalues of H are around one, the faster the convergence rate will be. To speed up the convergence, we may use the preconditioned conjugate gradient (PCG) method with preconditioner P_n (we will construct P_n in the coming section), i.e. we may solve the preconditioned system

$$(P_n^{-1}Q_n)^t(P_n^{-1}Q_n)\mathbf{x} = (P_n^{-1}Q_n)^t P_n^{-1}\mathbf{b}. \tag{2.6}$$

by the CG method. We note that if the preconditioned matrices $P_n^{-1}Q_n$ have a spectrum clustered around one, then the CG method, when applied to solving the system (2.6), will converge very fast.

In the following section we will construct preconditioners P_n for Q_n by exploiting the near-Toeplitz structure of Q_n. More precisely, we will write Q_n as the sum of a Toeplitz matrix T_n (a Toeplitz matrix is a matrix having constant diagonals) and a low rank matrix. The preconditioner P_n is an approximation to T_n. The construction is complicated by the fact that the generating function (the definition of the generating function will be discussed in the next section) g of T_n has a zero. We will see that P_n is the product of a circulant matrix (a circulant matrix is a particular class of Toeplitz matrices and we will discuss it shortly) and a band-Toeplitz matrix, where the band-Toeplitz matrix is used to cope with the zeros of g and the circulant matrix is to handle the remaining non-zero factors of g. We prove that if s, the number of servers, is independent of n, then for all sufficiently large n, P_n are invertible and that the preconditioned matrices $P_n^{-1}Q_n$ have a spectrum clustered around one. The preconditioned conjugate gradient method, when applied to the preconditioned system (2.6) will converge very fast. We will demonstrate this fast convergence by solving the system (2.4) for different choices of queuing parameters with different iterative methods and preconditioners. The numerical results in Section 2.4 show that the PCG method with our preconditioner P_n converges in a fixed number of steps independent of n and s, whereas the numbers of steps required by the Jacobi method increase like $O(n)$.

The outline of this chapter is as follows. In Section 2.2, we construct pre-conditioners for the linear system (2.4). We discuss the cost per iteration and memory requirement in Section 2.2.1. In Section 2.3, we give the convergence proof. Numerical examples are given in Section 2.4 to illustrate the fast convergence of the proposed method. Finally a summary is given in Section 2.5 to conclude the chapter.

2.2 Construction of Preconditioners

In this section, we discuss how to construct preconditioners for Q_n. We first give the definitions of *Toeplitz* and *circulant* matrices. Applications of Toeplitz systems can be found in [71]. A detailed survey on theory and applications of Toeplitz matrices and circulant preconditioning can be found in Chan and Ng [28]. An $n \times n$ matrix $T_n = (t_{i,j})$ is said to be Toeplitz if $t_{i,j} = t_{i-j}$, i.e. if T_n is constant along its diagonals. It is said to be circulant if its diagonals t_k further satisfy $t_{n-k} = t_k$ for $0 < k \leq n - 1$. We will use $\mathcal{P}_{2\pi}$ to denote the class of 2π-periodic continuous functions. For a given function $f \in \mathcal{P}_{2\pi}$, we let $\mathcal{T}_n[f]$ to be the $n \times n$ Toeplitz matrix with (j,k)th entry given by the $(j - k)$th Fourier coefficient of f. The function f is called the *generating function* of the sequence of Toeplitz matrices $\mathcal{T}_n[f]$.

There are quite a number of reasons why we consider circulant precondi-tioners. A circulant matrix C_n is characterized by its first column. When the first column

$$(c_0, c_1, \ldots, c_{n-1})^t$$

is given, one can write down the circulant matrix as follows:

$$C_n = \begin{pmatrix} c_0 & c_{n-1} & \cdots & \cdots & c_1 \\ c_1 & c_0 & c_{n-1} & \cdots & c_2 \\ \vdots & \ddots & \ddots & \ddots & \vdots \\ c_{n-2} & \ddots & \ddots & \ddots & c_{n-1} \\ c_{n-1} & c_{n-2} & \cdots & c_1 & c_0 \end{pmatrix}. \tag{2.7}$$

Secondly, a circulant matrix can be diagonalized by the discrete Fourier transform F_n, i.e. $C_n = F_n^* \Lambda_n F_n$, where

$$[F_n]_{j,k} = \frac{1}{\sqrt{n}} e^{\frac{(2jk\pi)i}{n}}, \quad j, k = 0, 1, \ldots, n - 1, \tag{2.8}$$

and F_n^* is the conjugate transpose of F_n and Λ_n is a diagonal matrix contain-ing the eigenvalues of C_n. Furthermore the eigenvalues of a circulant have analytical form; see Davis [62, p. 74] for instance. This is an advantage for the spectral analysis of the problem. The matrix vector multiplication of the forms $F_n \mathbf{y}$ and $F_n^* \mathbf{y}$ can be obtained in $O(n \log n)$ operations by the Fast

Fourier Transform (FFT). Thirdly, by exploiting the property of the discrete Fourier transform F_n, the eigenvalues of a circulant matrix and its inverse can also be obtained in $O(n \log n)$ operations. By using the following equations:

$$C_n^1 = C_n \mathbf{e_1} = F_n^* \Lambda_n F_n \mathbf{e_1} = F_n^* \Lambda_n \mathbf{1} = F_n^* \Lambda_n^1, \qquad (2.9)$$

the eigenvalues of C_n can be obtained in $O(n \log n)$ operations. Here

$$\mathbf{e_1} = (1, 0, \ldots, 0)^t,$$

C_n^1 is the first column of C_n,

$$\mathbf{1} = (1, 1, \ldots, 1)^t$$

and Λ_n^1 is the column vector containing all the eigenvalues of C_n. Hence solving a circulant linear system $C_n \mathbf{y} = \mathbf{b}$ requires only $O(n \log n)$ operations. Finally, we remark that, for a Toeplitz matrix T_n, the matrix vector multiplication of the form $T_n \mathbf{y}$ can be done in $O(n \log n)$ operations by embedding S_n into a $2n \times 2n$ circulant matrix; see Strang [109] for instance.

The idea of using circulant matrices as preconditioners for Toeplitz matrices has been studied extensively in recent years; see for instance [4, 8, 15, 21, 22, 29, 31, 81, 111]. For a given Toeplitz matrix T_n with diagonals $\{t_j\}_{-(n-1)}^{n-1}$, the T. Chan circulant preconditioner to T_n is defined to be the circulant matrix C_n which minimizes the Frobenius norm $||T_n - C_n||_F$ amongst all circulant matrices. The (i, j)th entry of C_n is given by c_{i-j} where

$$c_k = \begin{cases} \dfrac{(n-k)t_k + kt_{k-n}}{n}, & 0 \le k < n, \\ c_{n+k}, & 0 < -k < n, \end{cases} \qquad (2.10)$$

see T. Chan [34]. We note that the diagonals $\{c_j\}_{j=-(n-1)}^{n-1}$ and hence C_n can be obtained in $O(n)$ operations. For Toeplitz matrices generated by a function f, we will denote their T. Chan circulant preconditioners by $C_n[f]$. The following lemma shows that $C_n[f]$ is a good approximation to $T_n[f]$ as far as CG methods are concerned.

Lemma 2.2.1. Let $f \in \mathcal{P}_{2\pi}$. If f has no zeros, then for sufficiently large n, the eigenvalues of $C_n[f]$ are bounded uniformly from zero and the sequence of matrices $C_n[f]^{-1} T_n[f]$ has a spectrum clustered around one.

Proof. See Chan and Yeung [33].   □

Because of the good approximating properties, the T. Chan circulant preconditioner is useful in solving numerical elliptic partial differential equations [24, 32] and signal processing problems [30].

Returning to the queuing problems, we observe that the matrix Q_n given in (2.5) can be written as

$$Q_n = T_n + R_n, \tag{2.11}$$

where T_n is a Toeplitz matrix:

$$T_n = \begin{pmatrix} \lambda + s\mu & -s\mu & 0 & 0 & 0 & \cdots & 0 \\ -\lambda_1 & \lambda + s\mu & -s\mu & 0 & 0 & \cdots & 0 \\ -\lambda_2 & -\lambda_1 & \lambda + s\mu & \ddots & & \ddots & \vdots \\ \vdots & -\lambda_2 & \ddots & \ddots & -s\mu & \ddots & \\ \vdots & & \ddots & \ddots & \lambda + s\mu & \ddots & 0 \\ \vdots & & & \ddots & \ddots & \ddots & -s\mu \\ -\lambda_{n-1} & -\lambda_{n-2} & \cdots & \cdots & -\lambda_2 & -\lambda_1 & \lambda + s\mu \end{pmatrix}, \tag{2.12}$$

and R_n is a matrix of rank s. From (2.12), we see that $T_n = \mathcal{T}_n[g]$, where the generating function $g(\theta)$ is given by

$$g(\theta) = -s\mu e^{i\theta} + \lambda + s\mu - \sum_{k=1}^{\infty} \lambda_k e^{-ik\theta}. \tag{2.13}$$

We note that by (2.1), the Fourier coefficients of g are absolutely summable. In particular, by Weierstrass M-test, g is in $\mathcal{P}_{2\pi}$; see for instance Conway [61, p. 29].

Unfortunately, it is clear from (2.13) that $g(\theta)$ has a zero at $\theta = 0$ and hence Lemma 2.2.1 is not directly applicable. However, if we look at the real part of $g(\theta)$, we see that

$$\mathrm{Re}\{g(\theta)\} = -s\mu \cos\theta + \lambda + s\mu - \sum_{k=1}^{\infty} \lambda_k \cos(k\theta) \geq s\mu - s\mu \cos\theta. \tag{2.14}$$

Hence $g(\theta)$ has zero only at $\theta = 0$. Let us extend $g(\theta)$ to a function on the complex plane by letting $z = e^{-i\theta}$ and define

$$w(z) = -s\mu \frac{1}{z} + \lambda + s\mu - \sum_{k=1}^{\infty} \lambda_k z^k. \tag{2.15}$$

Then $z = 1$ is the only zero of w on the unit circle of the complex plane.

In the following, we assume that the power series $zw(z)$ will have a radius of convergence ρ greater than one. In particular, the function $zw(z)$ is analytic in the ball $B(0, \rho)$ with center at zero and radius ρ. We note that ρ is given by

$$\frac{1}{\rho} = \limsup |\lambda_j|^{1/j}, \tag{2.16}$$

see for instance Conway [61, p. 31]. By Weierstrass's factorization theorem [61, p. 79], we can write

$$w(z) = (1 - z)^\ell w_1(z),\qquad(2.17)$$

where ℓ is the order of the zero of $w(z)$ at $z = 1$ and is the least integer such that $w^{(\ell)}(1) \neq 0$. The function $w_1(z)$ will have no zeros on the unit circle and $zw_1(z)$ will be analytic in $B(0, \rho)$. We observe that by straightforward division,

$$zw(z) = (1 - z)\left\{-s\mu + \lambda z + \sum_{k=2}^{\infty}\left(\lambda - \sum_{j=1}^{k-1}\lambda_j\right)z^k\right\}.\qquad(2.18)$$

Hence if $\lambda > s\mu$, or if $\lambda = s\mu$ but $\lambda \neq \lambda_1$, then the factor in the braces is non-zero at $z = 1$ and therefore is equal to $w_1(z)$.

Using (2.17), we write

$$g(\theta) = (1 - e^{-i\theta})^\ell g_1(\theta),\quad \theta \in [-\pi, \pi],$$

where

$$g_1(\theta) = w_1(e^{-i\theta}).$$

Since $w_1(z)$ has no zeros on the unit circle, $g_1(\theta)$ has no zeros in $[-\pi, \pi]$. Moreover, since the power series $zw_1(z)$ is analytic in $B(0, \rho)$, with $\rho > 1$, the series $w_1(1)$ is absolutely convergent; see Conway [61, p. 31]. Therefore $g_1(\theta) \in \mathcal{P}_{2\pi}$ by the Weierstrass M-test [61, p. 29]. Let us look at the following example.

Example 2.2.1. If $\ell = 1$, then by (2.18)

$$g_1(\theta) = -s\mu e^{i\theta} + \lambda + \sum_{k=1}^{\infty}\left(\lambda - \sum_{j=1}^{k}\lambda_j\right)e^{-ik\theta}.\qquad(2.19)$$

In this case, the first n Fourier coefficients of g_1 can be computed recursively in $O(n)$ operations. Hence by using (2.10), $\mathcal{C}_n[g_1]$ can be constructed in $O(n)$ operations.

We now propose our preconditioner for Q_n. It is defined as

$$P_n = \mathcal{T}_n[(1 - e^{-i\theta})^\ell]\mathcal{C}_n[g_1].\qquad(2.20)$$

Notice that by definition, $\mathcal{T}_n[(1 - e^{-i\theta})^\ell]$ is a lower triangular matrix with half bandwidth ℓ. We remark that its diagonals can be obtained by binomial expansion and are therefore given explicitly by the Pascal triangle. Clearly the matrix has ones along its main diagonal and hence is invertible for all n. By the construction, $g_1(\theta)$ is a function in $\mathcal{P}_{2\pi}$ with no zeros in $[-\pi, \pi]$.

Therefore for n sufficiently large, we have by Lemma 2.2.1, $\mathcal{C}_n[g_1]$ is uniformly invertible, i.e. its eigenvalues are uniformly bounded away from zero. Hence we have proved:

Lemma 2.2.2. *If ρ defined in (2.16) is greater than one, then the preconditioner P_n defined in (2.20) is uniformly invertible for large n.*

Let us give an example to illustrate the construction of the preconditioners P_n.

Example 2.2.2. Let $0 < r < 1$. Suppose that $\lambda_i = r^i a$ for $i \geq 1$, where a is a constant such that $\sum_{i=1}^\infty \lambda_i = \lambda$. Then we have

$$g(\theta) = -\mu s e^{i\theta} + s\mu + \frac{ar}{1-r} - \sum_{k=1}^\infty a r^k e^{-ik\theta}.$$

Using (2.16), the radius of convergence $\rho = r^{-1} > 1$. By (2.18),

$$g(\theta) = (1 - e^{-i\theta}) \left\{ -s\mu e^{i\theta} + \frac{ar}{1-r} + \sum_{k=1}^\infty \frac{ar^{k+1}}{1-r} e^{-ik\theta} \right\}.$$

By direct checking, we see that if $ar \neq s\mu(1-r)^2$, the factor in the braces is non-zero at $\theta = 0$. In that case, $\ell = 1$ and

$$g_1(\theta) = -s\mu e^{i\theta} + \frac{ar}{1-r} + \sum_{k=1}^\infty \frac{ar^{k+1}}{1-r} e^{-ik\theta}. \tag{2.21}$$

2.2.1 Computational Cost

In each iteration of the PCG method, the main computational cost consists of solving a linear system

$$P_n \mathbf{y} = \mathbf{r} \tag{2.22}$$

and multiplying Q_n by some vector $\mathbf{r}$. Let us first consider the cost of solving (2.22). By definition,

$$P_n^{-1} = \mathcal{C}_n[g_1]^{-1} \mathcal{T}_n[(1 - e^{-i\theta})^\ell]^{-1}.$$

Given any vector $\mathbf{r}$, the matrix vector product

$$\mathcal{C}_n[g_1]^{-1} \mathbf{r}$$

can be done by using a Fast Fourier Transform (FFT) in $O(n \log n)$ operations. Since

$$\mathcal{T}_n[(1 - e^{-i\theta})^\ell]$$

is a lower triangular matrix with bandwidth ℓ, solving

$$\mathcal{T}_n[(1 - e^{-i\theta})^\ell]\mathbf{y} = \mathbf{r}$$

requires $O(\ell n)$ operations. Thus the system (2.22) can be solved in $O(n \log n) + O(\ell n)$ operations.

Next we consider the cost of computing $Q_n \mathbf{r}$. We will make use of the partitioning (2.11). Note that R_n in (2.11) is a matrix containing only $2s$ nonzero entries; we therefore need $O(s)$ operations for computing $R_n \mathbf{r}$. Since T_n is a Toeplitz matrix, $T_n \mathbf{r}$ can be computed in $O(n \log n)$ operations by embedding T_n into a $2n \times 2n$ circulant matrix; see Strang [109]. Hence $Q_n \mathbf{z}$ can be obtained in $O(n \log n)$ operations. Thus we conclude that the number of operations required for each iteration of the PCG method is of order $O(n \log n)$.

Finally, we consider the memory requirement. We note that besides some n-vectors, we only have to store the first column of the matrices $\mathcal{T}_n[(1 - e^{-i\theta})^\ell]$ and $\mathcal{C}_n[g_1]$ but not the whole matrices. Thus we need $O(n)$ memory for the PCG method.

2.3 Convergence Analysis

In this section, we prove that if the queuing system has s servers, where s is independent of n, then the preconditioned matrices $P_n^{-1} Q_n$ have a spectrum clustered around one. Hence the conjugate gradient methods will converge very fast when applied to solving the preconditioned normal system (2.6).

Lemma 2.3.1. *Let the radius of convergence ρ defined in (2.16) be greater than one. Then the sequence of matrices*

$$P_n^{-1}\mathcal{T}_n[g] = \mathcal{C}_n[g_1]^{-1}\mathcal{T}_n[(1 - e^{-i\theta})^\ell]^{-1}\mathcal{T}_n[g] \tag{2.23}$$

has a spectrum clustered around one for sufficiently large n.

Proof. The invertibility of the preconditioner

$$P_n = \mathcal{T}_n[(1 - e^{-i\theta})^\ell]\mathcal{C}_n[g_1]$$

is guaranteed by Lemma 2.2.2. Since

$$\mathcal{T}_n[(1 - e^{-i\theta})^\ell]$$

is a lower triangular Toeplitz of half bandwidth ℓ, it is straightforward to check that

$$\mathcal{T}_n[g] = \mathcal{T}_n[(1 - e^{-i\theta})^\ell]\mathcal{T}_n[g_1] + K_1,$$

where the rank of the matrix K_1 is no more than 2ℓ. Therefore

$$\begin{aligned}
\mathcal{C}_n[g_1]^{-1}\mathcal{T}_n[(1 - e^{-i\theta})^\ell]^{-1}\mathcal{T}_n[g] &= \mathcal{C}_n[g_1]^{-1}\mathcal{T}_n[(1 - e^{-i\theta})^\ell]^{-1} \\
&\quad \cdot (\mathcal{T}_n[(1 - e^{-i\theta})^\ell]\mathcal{T}_n[g_1] + K_1) \\
&= \mathcal{C}_n[g_1]^{-1}\mathcal{T}_n[g_1] + K_2,
\end{aligned}$$

where the rank of the matrix K_2 is no more than 2ℓ. Since g_1 has no zeros, by Lemma 2.2.1, the sequence of matrices $C_n[g_1]^{-1}\mathcal{T}_n[g_1]$ have a spectrum clustered around one for large n. It is now straightforward to show that the matrices in (2.23) also have a spectrum clustered around one.   □

Using the lemma above, we can now prove the clustering of eigenvalues for the preconditioned normal system (2.6).

Proposition 2.3.1. *Let the radius of convergence ρ defined in (2.16) be greater than one and the number of servers s in the queue be independent of the queue size n. Then the sequence of preconditioned matrices $P_n^{-1}Q_n$ has a spectrum clustered around one for large n.*

Proof. By (2.11) and (2.20),

$$P_n^{-1}Q_n = C_n[g_1]^{-1}\mathcal{T}_n[(1-e^{-i\theta})^\ell]^{-1}(\mathcal{T}_n[g] + R_n)$$
$$= C_n[g_1]^{-1}\mathcal{T}_n[(1-e^{-i\theta})^\ell]^{-1}\mathcal{T}_n[g] + K_3$$

where the rank of the matrix K_3 is no more than s. By Lemma 2.3.1, we see that $P_n^{-1}Q_n$ have a spectrum clustered around one.   □

It follows from standard convergence theory of the CG method that the method, when applied to solving the system (2.6), will converge very fast. In particular our numerical examples show that the method converges in finite number of steps independent of n.

2.4 Numerical Examples

In this section, we test the performance of our preconditioner P_n defined in (2.20). We compare the convergence rate of our method with the Jacobi method. As mentioned in Chapter 1, since Q_n is irreducibly diagonally dominant, both Jacobi and Gauss-Seidel methods converge when applied to solving the system (2.4). However, by using the partitioning of Q_n as in (2.11) and taking advantage of the Toeplitz matrix vector multiplication (see Section 2.2.1), we see that each iteration of the Jacobi method can be done in $O(n\log n)$ operations. However, this special property is not enjoyed by the Gauss-Seidel method, which will still require $O(n^2)$ operations per iteration. Thus in the numerical tests, we used only the Jacobi method in our comparisons.

For the CG type method, one can of course apply the ordinary CG method to the preconditioned normal system (2.6). But this requires the computation of $P_n^t\mathbf{y}$ and $Q_n^t\mathbf{y}$ for some vector $\mathbf{y}$ and will also square the condition number. Therefore in our experiments, we employed the CGS method to solve the preconditioned system $P_n^{-1}Q_n\mathbf{x} = P_n^{-1}\mathbf{b}$.

We compare the Jacobi method with the CGS method for two sets of queuing parameters:

1. $\lambda_j = \dfrac{1}{2^j}$, $j = 1, 2, \ldots$, and

2. $\lambda_j = \dfrac{90}{(\pi j)^4}$, $j = 1, 2, \ldots$.

We note that, in both cases, $\lambda = \sum_{k=1}^{\infty} \lambda_k = 1$. For each case, we tried three different choices of number of servers s: $s = 1$, 4, and n. The service rate μ is set to $\mu = \lambda/s$. We note that for the first case, the radius of convergence ρ defined in (2.16) is equal to 2 and the multiplicity of θ at zero is 1. Thus $g_1(\theta)$ is given by (2.21) with $a = 1$ and $r = 1/2$. This is the case covered by Proposition 2.3.1. For the second test case, we have $\rho = 1$ and hence it is a case not covered by Proposition 2.3.1. We will test it with $\ell = 1$ and $g_1(\theta)$ defined in (2.19).

The initial guess for both methods is $(1, 1, \ldots, 1)^t/n$. The stopping criterion for both the CGS and Jacobi method is

$$\|\mathbf{r}_k\|_2/\|\mathbf{r}_0\|_2 < 10^{-6},$$

where $\mathbf{r}_k$ is the residual at the kth iteration. All computations were done on a HP 715 workstation using MATLAB.

Tables 2.1 and 2.2 give the number of iterations required for convergence by using different methods and preconditioners. The J here means that the Jacobi method is used; I, P_n, and C_n mean that the CGS method is used without preconditioner, and with preconditioners P_n and $C_n[g]$ respectively. The symbol "**" signifies that the method does not converge in 5000 iterations.

Tables 2.1 and 2.2 show that the CGS method with our preconditioner P_n is the best choice, as is predicted by Proposition 2.3.1. We note from the tables that even when $s = n$ or for the second set of queuing parameters, which are cases not covered by Proposition 2.3.1, the performance of P_n is still the best amongst all the methods and preconditioners tested.

Table 2.1. Number of iterations for $\lambda_j = 2^{-j}$

n	$s = 1$				$s = 4$				$s = n$			
	I	P_n	$C_n[g]$	J	I	P_n	$C_n[g]$	J	I	P_n	$C_n[g]$	J
8	8	5	6	95	8	5	7	78	8	6	6	45
32	28	4	7	213	27	5	8	209	21	7	9	114
128	**	3	7	470	**	5	8	469	**	7	10	267
512	**	3	8	1331	**	5	8	1331	**	6	10	746

Table 2.2. Number of iterations for $\lambda_j = 90(\pi j)^{-4}$.

		$s = 1$				$s = 4$				$s = n$		
n	I	P_n	$\mathcal{C}_n[g]$	J	I	P_n	$\mathcal{C}_n[g]$	J	I	P_n	$\mathcal{C}_n[g]$	J
8	8	5	6	389	8	5	7	264	8	6	8	112
32	32	4	9	2253	32	6	10	2139	32	12	15	372
128	125	4	13	3874	124	5	15	3842	**	18	29	1005
512	**	3	21	**	**	5	21	**	**	17	38	3163

2.5 Summary and Discussion

A Markovian queuing system with batch arrival has been discussed in this chapter. The PCGS method was applied to solving the steady-state probability distribution with a Toeplitz-circulant type preconditioner. Numerical examples have been given to demonstrate the fast convergence rate of the proposed method.

Apart from computational issues, to apply the model to performance analysis, one has to estimate the system parameters even though the service rate μ can be controlled or known. For example, if the batch size follows a geometric distribution $(1 - p)p^{i-1} = 1, 2, \ldots$. The *Maximum likelihood* approach [75] is one possible choice. The method can be described as follows. Suppose that the following batch sizes are observed:

$$\{x_1, x_2, \ldots, x_m\}$$

then the maximum likelihood estimator $\tilde{p}$ is the one which maximizes the following probability:

$$P(x_1, x_2, \ldots, x_m, p) = \prod_{i=1}^{m}(1 - p)p^{x_i - 1}. \tag{2.24}$$

To obtain $\tilde{p}$ one may consider

$$\log(P(x_1, x_2, \ldots, x_m, p)) = m \log(1 - p) + \sum_{i=1}^{m}(x_i - 1) \log p.$$

Therefore we have

$$\frac{d \log(P(x_1, x_2, \ldots, x_m, p))}{dp} = \frac{-m}{1 - p} + \frac{1}{p} \sum_{i=1}^{m}(x_i - 1)$$

and

$$\frac{d^2 \log(P(x_1, x_2, \ldots, x_m, p))}{dp^2} = \frac{-m}{(1 - p)^2} - \frac{1}{p^2} \sum_{i=1}^{m}(x_i - 1) < 0.$$

By using standard calculus techniques one may obtain the maximum likelihood estimator for p as follows:

$$\tilde{p} = \frac{\displaystyle\sum_{i=1}^{m} x_i}{m + \displaystyle\sum_{i=1}^{m} x_i}. \tag{2.25}$$

Similarly, one can find the maximum likelihood estimator for the mean arrival rate λ and this is left as an exercise.

Exercises

2.1 Write down the generator A_n for $n = 3$ and $\lambda_i = 2^{-i}$. Solve the steady-state probability distribution in this case.

2.2 Show that the costs per iteration of the Jacobi method and the Gauss-Seidel method for solving the steady-state probability vector are $O(n \log n)$ and $O(n^2)$ respectively.

2.3 Given the first column of an $n \times n$ circulant matrix as follows:

$$(1, -1, 0, \ldots, 0)^t,$$

find all the eigenvalues of this matrix.

2.4 Explain why the system (2.4) can also be solved by Gaussian elimination in $O(n^2)$ operations with $O(n^2)$ memory.

2.5 The following are the inter-arrival times of a Poisson demands:

$$\{t_1, t_2, \ldots, t_m\}.$$

Find the maximum likelihood estimator for the mean arrival rate λ.

3. Circulant-Based Preconditioners for Queuing Systems with Markov-Modulated Poisson Process Inputs

3.1 Introduction to Markov-Modulated Poisson Processes

In this chapter, we study a queuing system which has a different arrival process compared with the queuing systems studied in Chapters 1 and 2. The arrival process is a *Markov-Modulated Poisson Process* (MMPP). An MMPP is a generalization of the Poisson process and is widely used as the input model of communication systems such as data traffic systems [72], Asynchronous Transfer Mode (ATM) networks [116].

Definition 3.1.1. *An MMPP is a Poisson process whose instantaneous rate is itself a stationary random process which varies according to an irreducible n-state Markov chain. We say that the MMPP is in phase k, $1 \le k \le n$, when the underlying Markov process is in state k, and in this case the arrivals occur according to a Poisson process of rate λ_k. The process is characterized by the generator matrix Q of the underlying Markov process and the rates $\lambda_1, \lambda_2, \ldots, \lambda_n$.*

Example 3.1.1. If n is 1, then the process is just a Poisson process.

Let us discuss a more complicated case when $n = 2$.

Example 3.1.2. Consider the servicing process of an unreliable exponential server. The server has two states: state 1 (the server is normal) and state 0 (the server is under repair). Assume that the normal time and the repair time of the machine are exponentially distributed with means σ_1^{-1} and σ_2^{-1} respectively. Suppose further that the server has service rate λ. Then the service process of the server is an MMPP of two states whose generator matrix and *diagonal rate matrix* are given by

$$Q_1 = \begin{pmatrix} \sigma_1 & -\sigma_2 \\ -\sigma_1 & \sigma_2 \end{pmatrix} \quad \text{and} \quad \Lambda_1 = \begin{pmatrix} \lambda & 0 \\ 0 & 0 \end{pmatrix} \tag{3.1}$$

respectively.

The superposition of two MMPPs is still an MMPP. Let us take a look at the following example.

Example 3.1.3. Consider a system that consists of two unreliable machines as described in Example 3.1.2. Let (i, j) denote the state that machine 1 is in state i and machine 2 is in state j $(i, j = 0, 1)$. We order the four states in the following sequence: $(1, 1), (1, 0), (0, 1), (0, 0)$. In State 1, both machines are normal and therefore the service rate of the system is 2λ. For State 2 and State 3, only one machine is normal and therefore the service rate is only λ. In State 4, both machines are under repair, so the service rate is of course zero:

$$
Q_2 = \begin{pmatrix}
\sigma_1 + \sigma_2 & -\sigma_2 & -\sigma_2 & 0 \\
-\sigma_1 & 2\sigma_2 & 0 & -\sigma_2 \\
-\sigma_2 & 0 & \sigma_1 + \sigma_2 & -\sigma_2 \\
0 & -\sigma_2 & -\sigma_1 & 2\sigma_2
\end{pmatrix}
\tag{3.2}
$$

and

$$
\Lambda_2 = \begin{pmatrix}
2\lambda & 0 & 0 & 0 \\
0 & \lambda & 0 & 0 \\
0 & 0 & \lambda & 0 \\
0 & 0 & 0 & 0
\end{pmatrix}.
\tag{3.3}
$$

By using tensor product defined in Chapter 1, the generator matrix Q_2 and the rate matrix Λ_2 can be written in the following compact form:

$$
Q_2 = I_2 \otimes Q_1 + Q_2 \otimes I_2 \quad \text{and} \quad \Lambda_2 = I_2 \otimes \Lambda_1 + \Lambda_1 \otimes I_2.
$$

Here I_2 is the 2×2 identity matrix and $\otimes$ denotes the Kronecker tensor product. It can be shown that the superposition of any n of the unreliable machines form an MMPP in the above sense.

In this chapter, we are going to present a numerical algorithm, PCGS, to solve the steady-state probability distributions of queuing systems with MMPP inputs. In Chapter 4, we will relate queuing systems with MMPP inputs to the production process in unreliable manufacturing systems. The remainder of the chapter is organized as follows. In Section 3.2, we discuss the queuing system (MMPP/M/s/$s + m$) arising in telecommunication systems. In Section 3.3, we construct preconditioners by taking circulant approximations of the tensor blocks of the generator matrices. In Section 3.4, we prove that the preconditioned systems have singular values clustered around one. In Section 3.5, we present a regularization method to handle the case when the generator matrix is ill-conditioned. The cost count of the proposed algorithm is given in Section 3.6. Numerical examples are given in Section 3.7 to illustrate the fast convergence rate of our method. Finally a summary is given in Section 3.8 to conclude the chapter.

3.2 The Queuing System

In this section, we present the queuing system (MMPP/M/s/$s + m$) arising in telecommunication networks; see Fig. 3.1 and Meier-Hellstern [90]. The

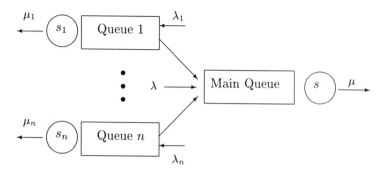

Fig. 3.1. The queuing system with MMPP inputs.

queuing system has $(q + 1)$ trunks and each trunk has m waiting spaces and s multiple exponential servers. The analysis of these queuing systems can be used to determine call congestion in tele-traffic networks with alternate routing; see Meier-Hellstern [90]. For interested readers, an introduction to telecommunication systems can be found in the book by Flood [67]. A call will overflow to other trunks if its first destination trunk is full and will be blocked from the system if all the trunks are full. The analysis of these queuing systems can be decomposed into the study of each trunk independently. For each trunk, the overflow from other trunks is modeled by a 2^q-state MMPP which is a superposition of q independent two-state MMPPs, i.e. each trunk is an (MMPP/M/s/$s + m$) queue. The generator matrices of these processes are $(s + m + 1)2^q \times (s + m + 1)2^q$ tri-diagonal block matrices with each diagonal block being a sum of tensor product of matrices. Our goal here is to solve for the steady-state probability distributions of the queues which are the normalized null vectors of the generator matrices.

Usually a classical iterative method, such as the Block Gauss-Seidel (BGS) method, is used to solve for the steady-state probability distribution. They are easy to implement, but their convergence rates are slow in general; see numerical results in Section 3.7. Here, we propose to use the Preconditioned Conjugate Gradient (PCG) method. Our preconditioners are constructed by taking a circulant approximation of the tensor blocks of the generator matrix. We prove that the preconditioned system has singular values clustered around one independent of the size of the waiting spaces m. Hence the conjugate gradient method will converge very fast when employed to solve the preconditioned system for large m. In fact, we prove that the number of iterations required for convergence grows at most like $O(\log m)$. Numerical examples are given in Section 3.7 to illustrate the fast convergence.

In order to construct the generator matrix of the queuing process, we first define the following queuing parameters:

(i) λ^{-1}, the mean arrival time of the exogenously originating calls,
(ii) μ^{-1}, the mean service time of each server,
(iii) s, the number of servers,
(iv) m, the number of waiting spaces in the queue,
(v) q, the number of overflow queues, and
(vi) (Q_j, Λ_j), $1 \leq j \leq q$, the parameters of the MMPP's modeling overflow parcels, where

$$Q_j = \begin{pmatrix} \sigma_{j1} & -\sigma_{j2} \\ -\sigma_{j1} & \sigma_{j2} \end{pmatrix} \quad \text{and} \quad \Lambda_j = \begin{pmatrix} \lambda_j & 0 \\ 0 & 0 \end{pmatrix}. \tag{3.4}$$

Here σ_{j1}, σ_{j2} and λ_j, $1 \leq j \leq q$, are positive MMPP parameters.

The input of the queue comes from the superposition of several independent MMPPs, which is still an MMPP and is parametrized by two $2^q \times 2^q$ matrices (Q, Γ). Here

$$Q = (Q_1 \otimes I_2 \otimes \cdots \otimes I_2) + (I_2 \otimes Q_2 \otimes I_2 \otimes \cdots \otimes I_2) + \cdots + (I_2 \otimes \cdots \otimes I_2 \otimes Q_q), \tag{3.5}$$

$$\Lambda = (\Lambda_1 \otimes I_2 \otimes \cdots \otimes I_2) + (I_2 \otimes \Lambda_2 \otimes I_2 \otimes \cdots \otimes I_2) + \cdots + (I_2 \otimes \cdots \otimes I_2 \otimes \Lambda_q) \tag{3.6}$$

and

$$\Gamma = \Lambda + \lambda I_{2^q}.$$

In the following, we are going to drop the subscript of the identity matrix I if the dimension of the matrix is clear from the context. We can regard our (MMPP/M/s/$s+m$) queue as a Markov process on the state space

$$\{(i, j) \mid 0 \leq i \leq s + m, 1 \leq j \leq 2^q\}.$$

The number i corresponds to the number of calls at the destination, while j corresponds to the state of the Markov process with generator matrix Q. Hence the generator matrix of the queuing process is given by the following $(s + m + 1)2^q \times (s + m + 1)2^q$ tri-diagonal block matrix A:

$$A = \begin{pmatrix} Q + \Gamma & -\mu I & & & & & 0 \\ -\Gamma & Q + \Gamma + \mu I & -2\mu I & & & & \\ & \ddots & \ddots & \ddots & & & \\ & & -\Gamma & Q + \Gamma + s\mu I & -s\mu I & & \\ & & & \ddots & \ddots & \ddots & \\ & & & & -\Gamma & Q + \Gamma + s\mu I & -s\mu I \\ 0 & & & & & -\Gamma & Q + s\mu I \end{pmatrix}. \tag{3.7}$$

For simplicity, let us write $n = (s + m + 1)2^q$. We recall that the steady-state probability distribution vector

$$\mathbf{p} = (p_1, p_2, \ldots, p_n)^t$$

is the solution to the matrix equation $A\mathbf{p} = \mathbf{0}$ with constraints

$$\sum_{i=1}^{n} p_i = 1$$

and

$$p_i \geq 0, \quad \text{for all } 1 \leq i \leq n.$$

The matrix A is irreducible, has zero column sums, positive diagonal entries and non-positive off-diagonal entries. Recall that from Chapter 1 (Perron-Frobenius theory) that the matrix A has a one-dimensional null space with a positive null-vector. Therefore the steady-state probability distribution vector $\mathbf{p}$ exists.

Many useful quantities of the telecommunication system such as the steady-state probability distribution of the number of calls at the destination, the blocking probability and the waiting time distribution can be obtained from the vector $\mathbf{p}$; see Flood [67] and Meier-Hellstern [90]. From Chapter 1, $\mathbf{p}$ can be obtained by normalizing the solution $\mathbf{x}$ of the non-singular system

$$G\mathbf{x} \equiv (A + \mathbf{e}_1 \mathbf{e}_1^t)\mathbf{x} = \mathbf{e}_1. \tag{3.8}$$

Here

$$\mathbf{e}_1 = (1, 0, \ldots, 0)^t$$

is the unit vector. We will solve the linear system (3.8) by using the CGS method. The convergence rate of CG type methods depends on the distribution of the singular values of the matrix G. The more clustered the singular values of G are, the faster the convergence rate will be. However, this is not the case for our matrix G and we will see in the numerical results in Section 3.7 that the convergence for the system (3.8) is slow. To speed up the convergence, a preconditioner is used. In essence, we solve, instead of (3.8), the preconditioned system

$$GC^{-1}\mathbf{w} = \mathbf{e}_1 \tag{3.9}$$

for $\mathbf{w}$ by CG type methods. Obviously, the solution $\mathbf{x}$ to (3.8) is given by $C^{-1}\mathbf{w}$. Recall that in Chapter 1, a good preconditioner C is a matrix such that it is easy to construct, the preconditioned matrix GC^{-1} has clustered singular values around one, and the preconditioned system $C\mathbf{y} = \mathbf{r}$ can be solved easily for any vector $\mathbf{r}$. We will show that our preconditioner satisfies these three criteria in the coming sections.

3.3 Construction of Circulant Preconditioners for the Queuing Systems

In this section, we discuss the construction of preconditioners for the linear system (3.9). The preconditioner C is constructed by exploiting the block structure of the generator matrix A in (3.7). Notice that the generator A can be written as the sum of tensor products:

$$A = I \otimes Q + B \otimes I + R \otimes \Lambda, \tag{3.10}$$

where B and R are $(s + m + 1) \times (s + m + 1)$ matrices given by

$$B = \begin{pmatrix} \lambda & -\mu & & & & & 0 \\ -\lambda & \lambda + \mu & -2\mu & & & & \\ & \ddots & \ddots & \ddots & & & \\ & & -\lambda & \lambda + s\mu & -s\mu & & \\ & & & \ddots & \ddots & \ddots & \\ & & & & -\lambda & \lambda + s\mu & -s\mu \\ 0 & & & & & -\lambda & s\mu \end{pmatrix}$$

and

$$R = \begin{pmatrix} 1 & & & & 0 \\ -1 & 1 & & & \\ & -1 & \ddots & & \\ & & \ddots & 1 & \\ 0 & & & -1 & 0 \end{pmatrix}.$$

For small s, we observe that B and R are close to the tri-diagonal Toeplitz matrices

$$\text{tridiag}[-\lambda, \lambda + s\mu, -s\mu] \quad \text{and} \quad \text{tridiag}[-1, 1, 0]$$

respectively. Recall that Toeplitz matrices are matrices having constant diagonals. Our preconditioner is then obtained by taking the *circulant approximation* of the matrices B and R, which are defined by $c(B)$ and $c(R)$ as follows:

$$c(B) = \begin{pmatrix} \lambda + s\mu & -s\mu & & & -\lambda \\ -\lambda & \lambda + s\mu & -s\mu & & \\ & \ddots & \ddots & \ddots & \\ & & -\lambda & \lambda + s\mu & -s\mu \\ -s\mu & & & -\lambda & \lambda + s\mu \end{pmatrix}, \tag{3.11}$$

and

$$c(R) = \begin{pmatrix} 1 & & & -1 \\ -1 & 1 & & \\ & -1 & \ddots & \\ & & \ddots & 1 & \\ 0 & & & -1 & 1 \end{pmatrix}. \tag{3.12}$$

We note that $c(B)$ and $c(R)$ are Strang's circulant approximations of the Toeplitz matrices

$$\text{tridiag}[-\lambda, \lambda + s\mu, -s\mu] \quad \text{and} \quad \text{tridiag}[-1, 1, 0]$$

respectively; see Chan [21]. Clearly, we have:

Lemma 3.3.1. $\text{rank}(B - c(B)) = s + 1$ and $\text{rank}(R - c(R)) = 1$.

Using the theory of circulant matrices; see Davis [62], we also have:

Lemma 3.3.2. *The matrices $c(B)$ and $c(R)$ can be diagonalized by the discrete Fourier transform matrix F, i.e.*

$$F^*(c(B))F = \Phi \quad \text{and} \quad F^*(c(R))F = \Psi$$

where both Φ and Ψ are diagonal matrices. The eigenvalues of $c(B)$ and $c(R)$ are given by

$$\phi_j = \lambda(1 - e^{\frac{-2\pi i(j-1)}{s+m+1}}) + s\mu(1 - e^{\frac{-2\pi i(j-1)(s+m)}{s+m+1}}), \quad j = 1, \ldots, s+m+1, \quad (3.13)$$

and

$$\psi_j = 1 - e^{\frac{-2\pi i(j-1)}{s+m+1}} \quad j = 1, \ldots, s + m + 1. \quad (3.14)$$

Thus, the matrices $c(B)$ and $c(R)$ can be inverted easily by using the Fast Fourier Transform (FFT).

We first approximate our matrix A in (3.10) (and hence G in (3.8)) by

$$D = I \otimes Q + c(B) \otimes I + c(R) \otimes \Lambda. \quad (3.15)$$

We observed that D is irreducible, has zero column sums, positive diagonal entries and non-positive off-diagonal entries. Hence D is singular and has a null space of dimension one. Moreover, D is unitarily similar to a diagonal block matrix:

$$(F^* \otimes I)D(F \otimes I) = I \otimes Q + \Phi \otimes I + \Psi \otimes \Lambda$$
$$= \text{Diag}(D_1, C_2, C_3, \ldots, C_{s+m+1}). \quad (3.16)$$

Here the blocks are

$$C_i = Q + \phi_i I + \psi_i \Lambda, \quad i = 2, \ldots, s + m + 1, \quad (3.17)$$

and $D_1 = Q$ with D_1 being the only singular block.

Let

$$C_1 = Q + \mathbf{e}_{2^q}\mathbf{e}_{2^q}^t, \quad (3.18)$$

where $\mathbf{e}_{2^q} = (0, \ldots, 0, 1)^t$ is a 2^q-vector. Since C_1 is irreducibly diagonal dominant with the last column being strictly diagonal dominant, it is non-singular. Our preconditioner C for the matrix G in (3.8) is defined as

$$C = (F \otimes I)\text{Diag}(C_1, C_2, \ldots, C_{s+m+1})(F^* \otimes I), \quad (3.19)$$

which is clearly non-singular.

3.4 Convergence Analysis

In this section, we study the convergence rate of our algorithm when m, the number of waiting spaces, is large. In the queuing systems considered in Meier-Hellstern [90], the number of waiting spaces m in each queue is much larger than the number of overflow queues q.

We prove that if all the queuing parameters λ, μ, s, q, and σ_{ij} are fixed independent of m, then the preconditioned system GC^{-1} in (3.9) has singular values clustered around one as m tends to infinity. Hence when CG type methods are applied to solving the preconditioned system (3.9), we expect fast convergence. Numerical examples are given in Section 3.7 to demonstrate our claim. We start the proof with the following lemma.

Lemma 3.4.1. *We have* $\operatorname{rank}(G - C) \leq (s + 2)2^q + 2$.

Proof. We note that by (3.8), we have

$$\operatorname{rank}(G - A) = 1.$$

From (3.10), (3.15), and Lemma 3.3.1, we see that

$$\operatorname{rank}(A - D) = (s + 2)2^q.$$

From (3.16), (3.18), and (3.19), we see that D and C differ by a rank one matrix. By using the fact that

$$\operatorname{rank}(S + T) \leq \operatorname{rank}(S) + \operatorname{rank}(T)$$

for any two matrices of same dimension (this is left as an exercise), we have

$$\begin{aligned}
\operatorname{rank}(G - C) &\leq \operatorname{rank}(G - A) + \operatorname{rank}(A - D) + \operatorname{rank}(D - C) \\
&= 1 + (s + 2)2^q + 1 \\
&= (s + 2)2^q + 2.
\end{aligned}$$

Hence the inequality is proved. $\qquad\square$

Proposition 3.4.1. *The preconditioned matrix* GC^{-1} *has at most* $2((s + 2)2^q + 2)$ *singular values not equal to one. Thus the matrix* GC^{-1} *has singular values clustered around one.*

Proof. We first note that

$$GC^{-1} = I + (G - C)C^{-1} \equiv I + L_1, \tag{3.20}$$

where $\operatorname{rank}(L_1) \leq (s + 2)2^q + 2$ by Lemma 3.4.1. Therefore

$$C^{-*}G^*GC^{-1} - I = L_1^*(I + L_1) + L_1, \tag{3.21}$$

is a matrix of rank at most $2((s + 2)2^q + 2)$. $\qquad\square$

The number of singular values of GC^{-1} that are distinct from one is a constant independent of m. In order to show fast convergence of preconditioned conjugate gradient (PCG) type methods with preconditioner C, one still needs an estimate of $\sigma_{\min}(GC^{-1})$, the smallest singular value of GC^{-1}. If $\sigma_{\min}(GC^{-1})$ is uniformly bounded away from zero independent of m, then the method converges in $O(1)$ iterations; and if $\sigma_{\min}(GC^{-1})$ decreases like $O(m^{-\alpha})$ for some $\alpha > 0$, then the method converges in at most $O(\log m)$ steps; see Van der Vorst [112] or Chan [19, Lemma 3.8.1].

In the following, we are going to show that even in the worst case where $\sigma_{\min}(GC^{-1})$ decreases in an order faster than $O(m^{-\alpha})$ for any $\alpha > 0$ (e.g. like $O(e^{-m})$), one can still have a fast convergence rate. We observe that in this case the matrix equation (3.9) is very ill-conditioned. Our trick is to consider a regularized equation of (3.9) as follows:

$$C^{-*}(G^*G + m^{-4-\beta}I)C^{-1}\mathbf{w} = C^{-*}G^*\mathbf{e}_n \qquad (3.22)$$

where β is any positive constant.

3.5 The Method of Regularization

In this section, we prove that the regularized preconditioned matrix

$$C^{-*}(G^*G + m^{-4-\beta}I)C^{-1}$$

has eigenvalues clustered around one and its smallest eigenvalue decreases at a rate no faster than $O(m^{-4-\beta})$. Hence PCG type methods will converge in at most $O(\log m)$ steps when applied to solve the preconditioned linear system (3.22). Moreover, we prove that the 2-norm of the error introduced by the regularization tends to zero at a rate of $O(m^{-\beta})$. In order to prove our claim, we have to get an estimate of the upper and lower bounds for $||C^{-1}||_2$. We begin our proof with the following lemma.

Lemma 3.5.1. *Given any matrix W, if the smallest eigenvalue of $W + W^*$, denoted by $\lambda_{\min}(W + W^*)$, satisfies*

$$\lambda_{\min}(W + W^*) \geq \delta > 0,$$

then $||W^{-1}||_2 \leq 2/\delta$.

Proof. For any arbitrary $\mathbf{x}$, using the Cauchy-Schwartz inequality, we have

$$\begin{aligned}
\delta||\mathbf{x}||_2^2 &\leq \lambda_{\min}(W + W^*)||\mathbf{x}||_2^2 \\
&\leq \mathbf{x}^*(W + W^*)\mathbf{x} \\
&= 2\mathbf{x}^*W\mathbf{x} \\
&\leq 2||\mathbf{x}||_2||W\mathbf{x}||_2.
\end{aligned}$$

Since $W\mathbf{x}$ is arbitrary, this implies that $||W^{-1}||_2 \leq 2/\delta$.   □

Lemma 3.5.2. *Let A and B be two $n \times n$ Hermitian matrices and let the eigenvalues $\lambda_i(A), \lambda_i(B)$, and $\lambda_i(A+B)$ be arranged in increasing order. Then for $k = 1, 2, \ldots, n$ we have*

$$\lambda_k(A) + \lambda_1(B) \leq \lambda_k(A + B) \leq \lambda_k(A) + \lambda_n(B).$$

Proof. See Horn and Johnson [76, p. 181]. □

Now we are ready to estimate $||C^{-1}||_2$.

Lemma 3.5.3. *Let the queuing parameters λ, μ, s, q, and σ_{ij} be independent of m. Then there exist positive constants τ_1 and τ_2 independent of m such that*

$$\tau_1 \leq ||C^{-1}||_2 \leq \tau_2 m^2.$$

Proof. We first prove the left hand side inequality. From (3.19), we see that C is unitarily similar to a diagonal block matrix. We therefore have

$$||C||_2 = \max\{||C_1||_2, ||C_2||_2, \ldots, ||C_{s+m+1}||_2\}.$$

Using (3.17), (3.13) and (3.14), it is straightforward to check that $||C_i||_1$ and $||C_i||_\infty$, $1 \leq i \leq s + m + 1$, are all bounded above by

$$\frac{1}{\tau_1} \equiv q\left(\max_j\{\sigma_{j1}\} + \max_j\{\sigma_{j2}\}\right) + 2(\lambda + s\mu + 1).$$

Using the inequality

$$|| \cdot ||_2 \leq \sqrt{|| \cdot ||_1 \cdot || \cdot ||_\infty},$$

we see that $||C_i||_2$, $i = 1, \ldots, s + m + 1$, are all bounded above by $1/\tau_1$. Thus $||C||_2 \leq 1/\tau_1$ and hence $\tau_1 \leq ||C^{-1}||_2$.

Next we prove the right hand side inequality. We note again by (3.19) that

$$||C^{-1}||_2 = \max\{||C_1^{-1}||_2, ||C_2^{-1}||_2, \ldots, ||C_{s+m+1}^{-1}||_2\}. \tag{3.23}$$

From (3.18), we can see that C_1 is a $2^q \times 2^q$ non-singular matrix with entries independent of m. Thus $||C_1^{-1}||_2$ is bounded independent of m. To obtain bounds for $||C_i^{-1}||_2$, $i = 2, \ldots, s + m + 1$, we first symmetrize the matrices. Define

$$\Sigma = \Sigma_1 \otimes \cdots \otimes \Sigma_q$$

where

$$\Sigma_j = \begin{pmatrix} 1 & 0 \\ 0 & \frac{\sigma_{j1}}{\sigma_{j2}} \end{pmatrix}, \qquad j = 1, 2, \ldots, q.$$

We observe that $||\Sigma||_2$ and $||\Sigma^{-1}||_2$ are bounded independent of m. By (3.4) and (3.5), we see that $Q\Sigma$ is a symmetric semi-definite matrix. Thus

$$C_i\Sigma = Q\Sigma + \phi_i\Sigma + \psi_i\Lambda\Sigma, \qquad i = 2, \ldots, s + m + 1,$$

are symmetric matrices too. By (3.14), we see that

$$(\psi_i \Lambda \Sigma + (\psi_i \Lambda \Sigma)^*), \quad i = 2, \ldots, s + m + 1,$$

are diagonal positive semi-definite matrices. Therefore we have

$$\lambda_{\min}(C_i \Sigma + (C_i \Sigma)^*) \geq \lambda_{\min}(\phi_i \Sigma + (\phi_i \Sigma)^*), \qquad i = 2, \ldots, s + m + 1. \quad (3.24)$$

From (3.13), we have

$$\lambda_{\min}(\phi_i \Sigma + (\phi_i \Sigma)^*) \geq \lambda \|\Sigma^{-1}\|_2^{-1} \sin^2\left(\frac{(i-1)\pi}{s+m+1}\right), \qquad i = 2, \ldots, s+m+1. \quad (3.25)$$

By making use of the following inequality (an exercise)

$$\sin\theta \geq \min\left\{\frac{2\theta}{\pi}, 2\left(1 - \frac{\theta}{\pi}\right)\right\}, \qquad \forall\theta \in [0, \pi], \quad (3.26)$$

we have

$$\lambda_{\min}(\phi_i \Sigma + (\phi_i \Sigma)^*) \geq \lambda \|\Sigma^{-1}\|_2^{-1} \min\left\{\frac{4(i-1)^2}{(s+m+1)^2}, 4\left(1 - \frac{i-1}{s+m+1}\right)^2\right\}$$

$$\geq \frac{4\lambda}{m^2}\|\Sigma^{-1}\|_2^{-1}, \qquad i = 2, \ldots, s+m+1.$$

By Weyl's theorem (Lemma 3.5.2), we then have

$$\lambda_{\min}(\phi_i \Sigma + (\phi_i \Sigma)^*) \geq \frac{\tau}{m^2}, \quad i = 2, \ldots, s+m+1, \quad (3.27)$$

where

$$\tau = 4\lambda \|\Sigma^{-1}\|_2^{-1}$$

is a positive constant independent of m.

Thus by (3.24), we get

$$\lambda_{\min}(C_i \Sigma + (C_i \Sigma)^*) \geq \frac{\tau}{m^2}, \quad i = 2, \ldots, s + m + 1. \quad (3.28)$$

Hence by Lemma 3.5.1, we have

$$\|\Sigma^{-1}C_i^{-1}\|_2 \leq \frac{2}{\tau}m^2, \qquad i = 2, \ldots, s + m + 1. \quad (3.29)$$

Therefore,

$$\|C_i^{-1}\|_2 \leq \|\Sigma\|_2\|\Sigma^{-1}C_i^{-1}\|_2 \leq \frac{2m^2}{\tau}\|\Sigma\|_2 \qquad i = 2, \ldots, s+m+1. \quad (3.30)$$

Since $||C_1^{-1}||_2$ is bounded above independent of m, we have

$$||C^{-1}||_2 \leq \max\left\{||C_1^{-1}||_2, \frac{2m^2}{\tau}||\Sigma||_2\right\} \equiv \tau_2 m^2, \qquad (3.31)$$

where τ_2 is a positive constant independent of m. Hence we have proved the lemma. $\square$

Proposition 3.5.1. *Let the queuing parameters* $\lambda, \mu, s, q,$ *and* σ_{ij} *be independent of* m. *Then for any positive* β, *the regularized preconditioned matrix*

$$C^{-*}(G^*G + m^{-4-\beta}I)C^{-1} \qquad (3.32)$$

has eigenvalues clustered around one and the smallest eigenvalue decreases at a rate no faster than $O(m^{-4-\beta})$. *Furthermore, the error introduced by the regularization is of the order* $O(m^{-\beta})$.

Proof. We note by Proposition 3.4.1 that

$$C^{-*}(G^*G + m^{-4-\beta}I)C^{-1} = I + L_2 + m^{-4-\beta}C^{-*}C^{-1}, \qquad (3.33)$$

where L_2 is a Hermitian matrix with

$$\mathrm{rank}(L_2) \leq 2((s+2)2^q + 2).$$

By Lemma 3.5.3, we have

$$\lim_{m\to\infty} m^{-4-\beta}||C^{-*}C^{-1}||_2 \leq \lim_{m\to\infty} m^{-\beta} = 0. \qquad (3.34)$$

Thus the regularized preconditioned matrix in (3.32) has eigenvalues clustered around one as m tends to infinity. The error introduced by the regularization method is given by $m^{-4-\beta}||C^{-*}C^{-1}||_2$ which by Lemma 3.5.3 tends to zero like $O(m^{-\beta})$.

As for the smallest eigenvalue of the regularized preconditioned matrix in (3.32), we note that

$$\min_{||\mathbf{x}||_2=1} \frac{\mathbf{x}^*(G^*G + m^{-4-\beta})\mathbf{x}}{\mathbf{x}^*C^*C\mathbf{x}} \geq \frac{\min_{||\mathbf{x}||_2=1}\mathbf{x}^*(G^*G + m^{-4-\beta})\mathbf{x}}{\max_{||\mathbf{x}||_2=1}\mathbf{x}^*C^*C\mathbf{x}} \qquad (3.35)$$
$$\geq \frac{\tau_1}{m^{4+\beta}},$$

where the right-most inequality follows from Lemma 3.5.3. We recall that τ_1 and β are positive constants independent of m. Hence the smallest eigenvalue of the regularized preconditioned matrix in (3.32) decreases no faster than $O(m^{-4-\beta})$. $\square$

Thus we conclude that PCG type methods applied to (3.22) with $\beta > 0$ will converge in at most $O(\log m)$ steps. To minimize the error introduced by the regularization method, one can choose a large β. Recall that regularization is required only when the smallest singular value of the matrix GC^{-1} in (3.9) tends to zero faster than $O(m^{-\alpha})$ for any $\alpha > 0$. In view of Lemma 3.5.3 (or cf. (3.35)), this can only happen when the smallest singular value of G has the same decaying rate. This will imply that the matrix G is very ill-conditioned. We note, however, that in all our numerical tests in Section 3.7, we found that there is no need to add the regularization.

3.6 Computational and Memory Cost Analysis

In this section, we derive the computational cost of the Preconditioned Conjugate Gradient (PCG) type method. We compare our PCG method with the Block Gauss-Seidel (BGS) method used in Meier-Hellstern [90]. We show that the cost for PCG type algorithms is

$$O(2^q(s+m+1)\log(s+m+1) + q(s+m+1)2^q).$$

The computational cost per iteration of the BGS method is $O((s+m+1)2^{2q})$; see Meier-Hellstern [90]. Thus PCG type methods require an extra $O(\log(s+m+1))$ of work per iteration compared with the BGS method. However, as we will soon see in the numerical examples of Section 3.7, the fast convergence of our method can more than compensate for this minor overhead in each iteration.

In PCG type algorithms for (3.9), the main cost per iteration is to compute the matrix vector multiplication of the form $GC^{-1}\mathbf{y}$ twice for some vector $\mathbf{y}$. By using the block tensor structure of A in (3.10), the multiplication of $G\mathbf{z}$ requires $(s+m+1)q2^q$ operations for any vector $\mathbf{z}$. By (3.19), we see that $C^{-1}\mathbf{y}$ is given by

$$(F \otimes I)\mathrm{Diag}(C_1^{-1}, C_2^{-1}, \ldots, C_{s+m+1}^{-1})(F^* \otimes I)\mathbf{y}.$$

It involves the matrix-vector multiplication of the form

$$(F^* \otimes I)\mathbf{z} \qquad \text{and} \qquad (F \otimes I)\mathbf{z}.$$

By using fast Fourier transforms, they can be obtained in

$$6(s+m+1)2^q \log(s+m+1)$$

operations. The vector

$$\mathrm{Diag}(C_1^{-1}, C_2^{-1}, \ldots, C_{s+m+1}^{-1})\mathbf{z} \tag{3.36}$$

can be obtained by solving $(s+m+1)$ linear systems involving the matrices C_i, $i = 1, \ldots, s+m+1$. Since each matrix is of size $2^q \times 2^q$, if Gaussian elimination is used, $O((s + m + 1)2^{3q})$ operations will be required. We now show that it can be reduced to $O((s + m + 1)q2^q)$ operations.

First we recall from the definitions of C_i, Q, and Λ in (3.17), (3.5), and (3.6) that

$$
\begin{aligned}
C_i = &((Q_1 + \phi_i I + \psi_i \Lambda_1) \otimes I \otimes \cdots \otimes I) \\
&+ (I \otimes (Q_2 + \phi_i I + \psi_i \Lambda_2) \otimes I \otimes \cdots \otimes I) \\
&+ \cdots + (I \otimes I \otimes \cdots \otimes (Q_q + \phi_i I + \psi_i \Lambda_q))
\end{aligned}
$$

where Q_j and Λ_j, $j = 1, \ldots, q$, are given in (3.4). By using Schur's triangularization theorem [76, p. 79], we can find 2×2 unitary matrices U_{ij} and lower triangular matrices L_{ij} such that

$$U_{ij}^*(Q_j + \phi_i I + \psi_i \Lambda_j)U_{ij} = L_{ij}, \qquad 1 \le i \le s + m + 1, \quad 1 \le j \le q. \quad (3.37)$$

For $i = 1, \ldots, s + m + 1$, define

$$U_i \equiv U_{i1} \otimes \cdots \otimes U_{iq}$$

and

$$L_i \equiv (L_{i1} \otimes I \otimes \cdots \otimes I) + (I \otimes L_{i2} \otimes I \otimes \cdots \otimes I) + \cdots + (I \otimes I \otimes \cdots \otimes I \otimes L_{iq}).$$

We see from (3.37) that

$$U_i^* C_i U_i = L_i, \quad 1 \le i \le s + m + 1.$$

Hence the vector $C_i^{-1} \mathbf{w}$ can be computed as $U_i L_i^{-1} U_i^* \mathbf{w}$.

The vector matrix multiplication of the form $U_j \mathbf{w}$ and $U_j^* \mathbf{w}$ can be done in $2(q2^q)$ operations by making use of the formula

$$U_j \mathbf{w} = (U_{1j} \otimes I \otimes \cdots \otimes I)(I \otimes U_{2j} \otimes I \otimes \cdots \otimes I) \cdots (I \otimes I \otimes \cdots \otimes I \otimes U_{qj})\mathbf{w}. \quad (3.38)$$

We note that the matrix L_i is a lower triangular matrix and each row of it has at most q non-zero entries. Hence $L_i^{-1} \mathbf{w}$ can be obtained in $q2^q$ operations. Thus for any vector $\mathbf{w}$, the vector $C_i^{-1} \mathbf{w}$ can be obtained in $3(q2^q)$ operations. Hence we conclude that the vector

$$\text{Diag}(C_1^{-1}, C_2^{-1}, \ldots, C_{s+m+1}^{-1})\mathbf{r}$$

can be computed in $3(s + m + 1)q2^q$ operations approximately.

In summary, each iteration of PCG type methods needs

$$2(6(s + m + 1)2^q \log(s + m + 1) + 4(s + m + 1)q2^q) \approx O(m \log m)$$

operations as compared to $O((s + m + 1)2^{2q}) \approx O(m)$ operations required by the BGS method. As we have proved in Section 3.4, PCG type methods will converge in at most $O(\log m)$ steps (see also the numerical results in Section

3.7), therefore the total complexity of our methods will be $O(m \log^2 m)$. As a comparison, the numerical results in Section 3.7 show that the number of iterations required for convergence for the BGS method increases linearly like $O(m)$. Therefore the total complexity of the BGS method is about $O(m^2)$ operations.

As for storage, PCG type methods and the BGS method require $O(2^q(s + m + 1))$ memory. Clearly, at least $O(2^q(s + m + 1))$ memory is required to store the approximated solution in each iteration.

3.7 Numerical Examples

In this section, we illustrate the fast convergence rate of our method by examples in queuing systems. The conjugate gradient squared (CGS) method is used to solve the preconditioned system (3.9). The stopping criterion for the CGS and BGS methods is set to be

$$\|A\mathbf{p}_k\|_2 \leq 10^{-12}, \tag{3.39}$$

where $\mathbf{p}_k$ is the computed steady-state probability distribution at the k-th iteration. The initial guess for both methods is the vector of all ones normalized such that its l_2-norm is equal to one. All the computations were done on an HP 712/80 workstation with MATLAB.

Let us give the numerical results for the queuing networks. We compare the numerical results of CGS, preconditioned CGS, and BGS methods for the number of overflow queues $q = 1, 2, 3, 4$, and the number of servers $s = 2$. The MMPP parameters are arbitrarily chosen to be

$$\sigma_{j1} = 2/3, \sigma_{j2} = 1/3, \quad j = 1, \ldots, q.$$

The other queuing parameters are given by

$$\mu = 2, \lambda = 1, \lambda_j = 1/q, j = 1, \ldots, q.$$

We recall that the size of the matrix is $(s+m+1)2^q \times (s+m+1)2^q$. The number of iterations required for convergence are given in Table 3.1. The symbols I, C, and BGS represent the methods used, namely, CGS without preconditioner, CGS with our preconditioner C in (3.19), and the Block Gauss-Seidel method respectively. Numbers of iterations greater than 2000 are signified by "**".

We see that the numbers are roughly constant independent of m for the CGS method with our preconditioner C. For the BGS method, the convergence rate is approximately linear in m. Recalling from Section 3.6 that the costs per iteration of the CGS method with preconditioning and of the BGS method are respectively $O(2^q(s+m+1) \log(s+m+1))$ and $O(2^{2q}(s+m+1))$ operations, we conclude that the total cost of obtaining the steady-state probability distribution vector for the CGS method with preconditioning is approximately $O(2^q(s + m + 1) \log(s + m + 1))$ operations, while for the BGS method, it is approximately $O(2^{2q}m(s + m + 1))$ operations.

Table 3.1. Number of iterations for convergence ($s = 2$)

	$q = 1$			$q = 2$			$q = 3$			$q = 4$		
m	I	C	BGS	I	C	BGS	I	C	BGS	I	C	BGS
16	36	7	130	36	9	112	38	12	107	40	13	110
64	**	7	242	**	9	207	**	12	213	**	13	199
256	**	8	601	**	10	549	**	12	582	**	14	530
1024	**	8	**	**	10	**	**	12	**	**	14	**

3.8 Summary and Discussion

A Markovian queuing system with MMPP inputs has been discussed in this chapter. The PCGS method has been applied to solve the steady-state probability distribution with a circulant-based preconditioner. Numerical examples have been given to demonstrate the fast convergence rate of the proposed method. The preconditioning algorithm can be further speeded up by using parallel computing techniques because (3.36) can be solved in parallel. A regularization technique has been developed in case the generator matrix is very ill-conditioned. But as demonstrated in the numerical examples and many other tested numerical examples, the regularization is not necessary. The preconditioning techniques developed here can also be applied to solve a large class of networks, namely Stochastic Automata Networks (SANs) [2, 12, 13, 14]; see for instance [26, 27, 57, 43, 96, 108].

Exercises

3.1 Consider a service system which consists of n unreliable machines as described in the beginning of this chapter. Show that the service process of the system of machines is an MMPP. What are the generator matrix and the rate matrix of the system?

3.2 Consider the system consists of two unreliable machines as described in Example 3.1.3. Show that the steady-state probability distribution is given by

$$\left(\frac{\sigma_2}{\sigma_1 + \sigma_2}, \frac{\sigma_1}{\sigma_1 + \sigma_2} \right)^t \otimes \left(\frac{\sigma_2}{\sigma_1 + \sigma_2}, \frac{\sigma_1}{\sigma_1 + \sigma_2} \right)^t.$$

If the operating cost and repairing cost per unit time of a machine are C and R respectively, find the expected running cost of the system in terms of C, R, σ_1, and σ_2. Show that the expected running cost is $C + R$ if $\sigma_1 = \sigma_2$.

3.3 Prove Lemma (3.3.2). Discuss the asymptotic behavior of ϕ_2 and ν_2 when m is large.

3.4 Show that for any vector $\mathbf{x}$ we have

$$||\mathbf{x}||_2^2 \leq ||\mathbf{x}||_\infty ||\mathbf{x}||_1.$$

3.5 Let S and T be two $l \times k$ matrices, explain why

$$\operatorname{rank}\begin{pmatrix} S \\ T \end{pmatrix} \leq \operatorname{rank}(S) + \operatorname{rank}(T).$$

Hence show that
$$\operatorname{rnk}(S + T) \leq \operatorname{rank}(S) + \operatorname{rnk}(T).$$

3.6 To prove the inequality (3.26), we define

$$f(\theta) = \sin\theta - \frac{2\theta}{\pi}.$$

Show that $f(0) = 0$ and $f'(\theta) \geq 0$ for $0 \leq \theta \leq \pi/2$. Deduce that

$$f(\theta) \geq 0 \quad \text{for} \quad 0 \leq \theta \leq \pi/2$$

and hence show that

$$\sin\theta \geq \min\left\{ \frac{2\theta}{\pi}, 2\left(1 - \frac{\theta}{\pi}\right) \right\}, \qquad \forall\theta \in [0, \pi].$$

4. Application of MMPP to Manufacturing Systems of Multiple Unreliable Machines

4.1 Introduction

In this chapter, we study an application of the queuing system discussed in Chapter 3. We apply the queuing system to modeling a manufacturing system of multiple failure prone machines under *Hedging Point Production* (HPP) policy. In a queuing system, there are servers, customers, and waiting spaces. To model a manufacturing system by a queuing system, one may regard a server as a machine. Depending on the type of manufacturing system, *make-to-order* or *order-to-make*, the customers can be regarded as the inventory of product or the jobs to be processed respectively; see Buzacott and Shanthikumar [17]. In our context here, we only discuss the latter case. In a make-to-order system, the machine system produces only when there is an arrived order. But in an order-to-make system, a certain amount of inventory is kept to cope with the fluctuation of demand and therefore production control is necessary. Here we employ the HPP policy as the production control; the definition of the policy will be discussed shortly.

The rest of the chapter is organized as follows. An introduction to HPP policy is given in Section 4.2. In Section 4.3, we applied the method derived in Chapter 3 to the production planning of manufacturing systems with multiple unreliable machines. Numerical examples are given in Section 4.4 to illustrate the fast convergence rate of our method. A summary will be given in Section 4.5 to conclude this chapter.

4.2 The Hedging Point Production Policy

An HPP policy is characterized by a number h: the machines keep on producing at their maximum possible production rates if the inventory level is less than h and the production is stopped when the inventory level reaches h. In fact, the optimal value of h is the best amount of inventory to be kept in the system so as to hedge against the mentioned uncertainties. When the optimal policy is a zero-inventory policy (i.e. the hedging point h is zero), then the policy matches the just-in-time (JIT) policy. Thus one may regard the HPP policy as a generalization of JIT. The JIT policies have strongly been favored in real-life production systems for process discipline reasons even when they are not optimal. By using the JIT policy, the Toyota company can manage to reduce the work-in-process and the cycle time in the presence of the stochastic situations mentioned above; see Monden [92]. We focus on finding optimal HPP policies for the captured manufacturing systems.

Example 4.2.1. To understand the HPP policy with hedging point being h, let us consider a system of one reliable machine producing at a rate of λ, and at the same time we let the demand rate be μ. We assume that the demand rate μ is less than the product rate λ. We allow a maximum backlog of one in the system. Excessive demand will then be rejected. The following is the Markov chain when the inventory level is used to describe the state of the system (see Fig. 4.1), where "-1" represents the backlog state. Using the

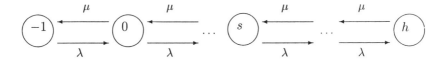

Fig. 4.1. The Markov chain for the reliable machine system.

result in Chapter 1, one can get the steady-state probability distribution of the inventory levels as follows:

$$p_i = \frac{\rho - 1}{\rho^{n+2} - 1} \rho^{i+1}, \quad i = -1, 0, \ldots, h.$$

Here p_i is the steady-state probability that the inventory level is i and $\rho = \lambda/\mu > 1$. Therefore the expected inventory level is given by

$$\sum_{i=1}^{h} i p_i = \frac{\rho^2}{\rho^{h+2} - 1} \left(\frac{h\rho^{h+1} - (h + 1)\rho^h + 1}{\rho - 1} \right). \tag{4.1}$$

Suppose that $\mu = 1$, the unit inventory cost is 1, and the unit backlog cost is 100, then the average running cost of the system is given by

$$C(h) = 100 \cdot p_{-1} + 1 \cdot \sum_{i=1}^{h} i p_i.$$

Table 4.1 shows the optimal h which minimizes $C(h)$ for different ρ. We remark that from the computations, zero-inventory is optimal when $\rho > 50$. This means that under the specified system parameters, if the production rate is at least 50 times of the demand rate then the optimal HPP policy is to keep nothing in the buffer!

Table 4.1. The optimal hedging point h^* for different ρ

ρ^{-1}	0.02	0.1	0.2	0.4	0.6	0.8	0.9
h^*	0	1	2	3	5	8	13

4.3 Manufacturing Systems of Multiple Failure Prone Machines

In this section, we consider manufacturing systems of q multiple unreliable machines producing one type of product with arrival of demand being a Poisson process. The machines are unreliable; when a machine breaks down it is subject to an exponential repairing process. The normal time of a machine and the processing time of one unit of product are exponentially distributed. Finished product is stored in a buffer; see Fig. 4.2. Moreover there is an inventory cost for holding each unit of product and a shortfall cost for each unit of backlog. Usually a proper inventory is stored to hedge against un-

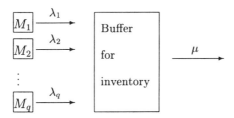

Fig. 4.2. The manufacturing system of unreliable machines.

certain situations such as breakdown of machines and shortfall of products; see for instance Akella and Kumar [3]. It is well-known that the HPP policy is optimal for one-machine manufacturing systems in some simple situations [3, 11, 77, 79]. In those works, the discrete inventory level of the product is approximated by a continuous fluid flow model. Analytic optimal control is found to be threshold type by solving a pair of Hamilton-Jacobi-Bellman equations. The control is optimal in the sense that it minimizes the average (or discounted) running cost of the manufacturing systems. For two-machine flowshops, HPP policies are no longer optimal but are near-optimal; see Sethi et al. [103]. Here we focus on finding the optimal hedging point for the manufacturing systems under consideration.

It should be noted that in [3, 11, 56, 77, 78, 79, 80, 104, 105, 115], only one machine is considered and the machine has only two states, normal and repairing. Here we consider a system consisting of q multiple unreliable machines. The production process of the machines is then an MMPP. The states of the machines and the inventory level can be modeled as an irreducible continuous time Markov chain. For different values of the hedging point h, the average running cost $C(h)$ can be written in terms of the steady-state probability distribution of the Markov chain. Therefore the optimal hedging point can be obtained by varying different values of h. Let us define the following parameters for the manufacturing systems as follows:

(i) q, the number of machines,

(ii) σ_{j1}^{-1}, the mean up time of the machine j, $j = 1, \ldots, q$,

(iii) σ_{j2}^{-1}, the mean repair time for the machine j, $j = 1, \ldots, q$,

(iv) λ_j^{-1}, the mean processing time for one unit of product on machine j, $j = 1, \ldots, q$,

(v) μ^{-1}, the mean inter-arrival time of demand,

(vi) h, the hedging point, and

(vii) g, the maximum allowable backlog.

For each machine j, $j = 1, \ldots, q$, let Q_j be the generator matrix of the machine states and Λ_j be the corresponding production rate matrix. Here

$$Q_j = \begin{pmatrix} \sigma_{j1} & -\sigma_{j2} \\ -\sigma_{j1} & \sigma_{j2} \end{pmatrix} \quad \text{and} \quad \Lambda_j = \begin{pmatrix} \lambda_j & 0 \\ 0 & 0 \end{pmatrix}$$

(cf. (3.4)). Each machine has two states, either "normal" or "repairing". Since there are q machines, there are 2^q states for the system of machines. We denote the set of machines states by Ω. The superposition of the q machines forms an MMPP and is characterized by the following $2^q \times 2^q$ generator matrix:

$$Q = (Q_1 \otimes I_2 \otimes \cdots \otimes I_2) + (I_2 \otimes Q_2 \otimes I_2 \otimes \cdots \otimes I_2) + \cdots + (I_2 \otimes \cdots \otimes I_2 \otimes Q_q) \quad (4.2)$$

(cf. (3.5)). The corresponding production rate matrix is given by

$$\Lambda = (\Lambda_1 \otimes I_2 \otimes \cdots \otimes I_2) + (I_2 \otimes \Lambda_2 \otimes I_2 \otimes \cdots \otimes I_2) + \cdots + (I_2 \otimes \cdots \otimes I_2 \otimes \Lambda_q) \quad (4.3)$$

(cf. (3.6)).

We note that the generator matrix for the machine-inventory system is a particular case of the queuing systems discussed in Chapter 3, with the queue size m being the size of the inventory, which in practice can easily go up to the thousands. Therefore our numerical method developed for the queuing networks above is very suitable for solving the steady-state probability distribution for these processes. Given a hedging point, the average running cost of the machine-inventory system can be written in terms of the steady-state probability distribution. Hence the optimal hedging point can also be obtained.

We let $\alpha(t)$ be the state of the system of machines at time t. Therefore $\alpha(t)$ has 2^q possible states. The inventory level takes an integer value in $[-g, h]$, because we allow maximum backlog of g and the hedging point is h. Here negative inventory means backlog. We let $x(t)$ be the inventory level at time t. The machines inventory process

$$\{(\alpha(t), x(t)), t \geq 0\}$$

forms an irreducible continuous time Markov chain in the state space

$$\{(\alpha, x) \mid \alpha \in \Omega, \; x = -g, \ldots, 0, \ldots, h\}.$$

If we order the states of the machine-inventory process lexicographically, we get the following $(h + g + 1)2^q \times (h + g + 1)2^q$ generator matrix H for the machine-inventory system:

$$H = \begin{pmatrix} Q + \Lambda & -\mu I & & & & 0 \\ -\Lambda & Q + \Lambda + \mu I & -\mu I & & & \\ & \ddots & \ddots & \ddots & & \\ & & -\Lambda & Q + \Lambda + \mu I & -\mu I & \\ & & & \ddots & \ddots & \ddots \\ & & & & -\Lambda & Q + \Lambda + \mu I & -\mu I \\ 0 & & & & & -\Lambda & Q + \mu I \end{pmatrix},$$

(4.4)

where I is the $2^m \times 2^m$ identity matrix. Clearly, the matrix H in (4.4) has the same tensor block structure as that of the generator matrix A in (3.7) in Chapter 3. In fact, H is a particular case of A with

$$s = 1, \quad \lambda = 0 \quad \text{and} \quad m = h + g - 1.$$

Therefore the techniques and algorithms developed in the previous sections can be used to obtain the steady-state distribution of the process efficiently. Numerical results are given in Section 4.3 to illustrate the fast convergence.

One of the important quantities, the average running cost of the machine-inventory system, can be written in terms of its steady-state probability distribution. Let

$$p(\alpha, x) = \lim_{t \to \infty} \text{Prob} \{\alpha(t) = \alpha, x(t) = x\} \tag{4.5}$$

be the steady-state probability distribution, and let

$$p_j = \sum_{k \in \Omega} p(k, j), \quad j = -g, -(g-1), \ldots, 0, \ldots, h, \tag{4.6}$$

be the steady-state probability distribution of the inventory level of the system. The average running cost for the machine-inventory system is then given by

$$C(h) = c_I \sum_{j=1}^{h} j p_j - c_B \sum_{j=-g}^{-1} j p_j, \quad 0 \leq h \leq b, \tag{4.7}$$

where

(a) c_I is the inventory cost per unit of product,

(b) c_B is the backlog cost per unit of product,

(c) b is the maximum inventory capacity.

Hence once p_j are given, we can easily find an h^* which minimizes the average running cost function $C(h)$ by evaluating $C(h)$ for all $0 \leq h \leq b$.

We remark that our method can be generalized to handle the case where each machine has the Erlangian distribution of l phases. Suppose the mean times of repair for machine $j, j = 1, \ldots, q$, are the same in each phase and are equal to $1/\sigma_{j2}$. In this case, the generator matrix for the machine-inventory system can be obtained by replacing the generator matrix of the machine states and its corresponding production rate matrix by $\bar{Q}_j$ and $\bar{\Lambda}_j$ respectively, where

$$
\bar{Q}_j = \begin{pmatrix}
\sigma_{j1} & & & & -\sigma_{j2} \\
-\sigma_{j1} & \sigma_{j2} & & & \\
& -\sigma_{j2} & \sigma_{j2} & & \\
& & \ddots & \ddots & \\
0 & & & -\sigma_{j2} & \sigma_{j2}
\end{pmatrix}
\tag{4.8}
$$

and

$$
\bar{\Lambda}_j = \begin{pmatrix}
\lambda_i & & & & 0 \\
& 0 & & & \\
& & 0 & & \\
& & & \ddots & \\
0 & & & & 0
\end{pmatrix}.
\tag{4.9}
$$

Hence we see that the techniques and algorithms developed previously can be applied to this case too.

4.4 Numerical Examples

In this section, we test our algorithm for the failure prone manufacturing systems. We illustrate the fast convergence rate of our method by examples in manufacturing systems. The stopping criterion for CGS and BGS methods is set to be

$$
||A\mathbf{p}_k||_2 \leq 10^{-12},
\tag{4.10}
$$

where $\mathbf{p}_k$ is the computed steady-state probability distribution at the kth iteration. We assume that all q machines are identical, and in each month (four weeks), each machine breaks down once on average. The mean repairing time for a machine is one week. Therefore we have

$$
\sigma_{j1} = 1/3, \quad \sigma_{j2} = 1, \quad j = 1, \ldots, q.
$$

The mean time for the arrival of demand is $1/5$ week and the mean time for the machine system to produce one unit of product is one day, therefore we have

$$\mu = 5 \quad \text{and} \quad \lambda_j = 7/q, \quad j = 1, \ldots, q.$$

In Table 4.2, we give the number of iterations required for convergence for all three methods. Recall that the symbols I, C, and BGS represent the methods used, namely, CGS without preconditioner, CGS with our preconditioner C in (3.19) and the Block Gauss-Seidel (BGS) method respectively. Numbers of iterations greater than 2000 are signified by "**". As in the queuing systems of previous chapter, we see also that the numbers are roughly constant independent of $(g+h)$ for the CGS method with our preconditioner C. For the BGS method, the convergence rate is again approximately linear in $(g+h)$. Finally, we consider examples of finding the optimal hedging

Table 4.2. Number of iterations for convergence

$g+h$	$q = 1$			$q = 2$			$q = 3$			$q = 4$		
	I	C	BGS	I	C	BGS	I	C	BGS	I	C	BGS
16	52	6	565	54	7	603	60	8	601	63	9	685
64	**	6	**	**	8	**	**	9	**	**	10	**
256	**	8	**	**	9	**	**	9	**	**	10	**
1024	**	8	**	**	9	**	**	9	**	**	11	**

point h^*. We keep the values of the machine parameters the same as in the manufacturing system example above, except that we set $q = 4$ and $g = 50$. Moreover, the inventory cost c_I and backlog cost c_B per unit of product are 50 and 2000 respectively, the maximum inventory capacity b is 200, see (4.7). In Table 4.3, we give the optimal pair of values $(h^*, C(h^*))$, the optimal hedging point h^* and its corresponding average running cost per week $C(h^*)$ for different values of λ_i and μ.

Table 4.3. The optimal $(h^*, C(h^*))$ for different λ_i and μ

	$\mu = 1$	$\mu = 2$	$\mu = 3$
$\lambda_i = 1$	(3, 181)	(10, 533)	(200, 14549)
$\lambda_i = 1.5$	(2, 128)	(5, 270)	(11, 576)

4.5 Summary and Discussion

Markovian queuing system with MMPP inputs has been applied to model the production process of manufacturing systems of multiple unreliable machines. HPP policy was employed as the production control and the problem was formulated as a Markov chain model. Given a hedging point h, the average running cost of the system can be written down in terms of the steady-state probability distribution. The numerical algorithm developed in Chapter 3 is applied to solving the steady-state probability distribution of the system. The objective was to find the optimal HPP policy (hedging point) such that the average running cost of the system is minimized. It is also interesting to extend our model to a more general demand process; see Ching [47].

Exercises

4.1 Consider the one-machine system under HPP policy in Example 4.2.1. Suppose there is no limit in backlog and the production rate λ is greater than the demand rate μ. Let h be the hedging point.
(a) Show that the steady-state probability distribution of the inventory levels exists and has the analytical solution:

$$q(i) = (1 - p)p^{n-i}, \quad i = n, n - 1, n - 2, \ldots$$

where $p = \mu/\lambda$ and $q(i)$ is the steady-state probability that the inventory level is i.
(b) Suppose I is the unit inventory cost and B is the unit backlog cost. Show that the expected running cost of the system (sum of the inventory cost and the backlog cost) can be written down as follows:

$$E(h) = I \underbrace{\sum_{i=0}^{h}(n - i)(1 - p)p^i}_{\text{inventory cost}} + B \underbrace{\sum_{i=h+1}^{\infty}(i - h)(1 - p)p^i}_{\text{backlog cost}}.$$

(c) Show that the expected running cost $E(h)$ is minimized if the hedging point h is chosen such that

$$p^{h+1} \leq \frac{I}{I + B} \leq p^h.$$

Hint: note that

$$E(h - 1) - E(h) = B - (I + B)(1 - p)\sum_{i=0}^{h-1}p^i$$

and

$$E(h+1) - E(h) = -B + (I+B)(1-p) \sum_{i=0}^{h} p^i.$$

(d) Show also that when $p \leq I(I+B)^{-1}$ we have

$$E(0) \leq E(1) \leq E(2) \leq \cdots.$$

and therefore the zero-inventory policy (i.e. $h = 0$) or a just-in-time policy can be optimal in this case.

4.2 Write down the generator matrix for the manufacturing system when the generator matrix of the machine states and its corresponding production rate matrix take the forms in (4.8).

5. Manufacturing Systems with Batch Arrivals

5.1 Introduction

In this chapter, we study an unreliable one-machine manufacturing system. The manufacturing system consists of one machine producing one type of product; see Fig. 5.1. The machine M_1 is unreliable and is subject to random breakdowns and repairs. Finished product is stored in the buffer B_1. The processing time for one unit of product and the normal time of the machine are exponentially distributed. The demand of the product is modeled as finite batch arrival with the inter-arrival time being exponentially distributed. The batch size distribution is stationary (independent of time). If the arrival demand exceeds the maximum allowable backlog, the system will accept part of the demand (up to the maximum allowable backlog). The demand process is similar to that in Chapter 2. HPP policy is employed as the production control. We focus on finding optimal hedging point policies for the captured manufacturing systems.

Very often, instead of one task, a repairing process requires a sequence of tasks. Here, we will consider this general situation for the repairing time. When the machine breaks down, it is subject to a sequence of l repairing phases and in each phase the repairing time is exponentially distributed, which is an Erlangian distribution. In a manufacturing environment under total quality management practice, repairing of machines, examination of failure factors and review on working procedures are necessary steps; see Flynn et al. [35, 68]. Let us first define the following parameters for the

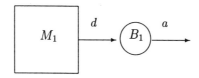

Fig. 5.1. Manufacturing system of a unreliable machine.

model.

(i) λ^{-1}, the mean up time of the machine,

(ii) d^{-1}, the mean processing time for one unit of product,

(iii) μ_i^{-1}, the mean repair time for the machine in phase i, $i = 1, 2, \ldots, l$,

(iv) a^{-1}, the mean inter-arrival time of demand,

(v) q_i, the probability that the arrival demand requires i units of product,
$i = 1, 2, \ldots, k$ and $\sum_{i=1}^{k} q_i = 1$,

(vi) m, the maximum allowable backlog,

(vii) b, the maximum inventory capacity,

(viii) h, the hedging point,

(ix) $c_I > 0$, the unit inventory cost,

(x) $c_B > 0$, the unit backlog cost.

Recall in Chapter 4, proper positive inventory level is maintained in the manufacturing system to hedge against the uncertainties in demand and breakdowns of the machine. Recall that an HPP policy is characterized by a number h. The machine keeps on producing products if the inventory level is less than h and the machine state is "normal", otherwise the machine is shut down or under repair. The machine states and the inventory levels are modeled as Markovian processes. It turns out that the combined machine-inventory process is an irreducible continuous time Markov chain. We give the generator matrix for the whole process, and then we employ the Pre-conditioned Conjugate Gradient (PCG) method to compute the steady-state probability distribution. Preconditioners are constructed by taking a circulant approximation of the near-Toeplitz structure of the generator matrix of the machine-inventory system. We prove that if the parameters a, d, λ, l, k, q_i, and μ_i are fixed independent of $n = m + h + 1$, then the preconditioned linear system has singular values clustered around one as n tends to infinity. Hence the Conjugate Gradient (CG) type methods will converge very fast when applied to solving the preconditioned linear system. Numerical examples are given in Section 5.6 to verify our claim. The average running cost for the system can then be written in terms of the steady-state probability distribution, and therefore the optimal hedging point can be obtained by varying different values of h.

The rest of this chapter is organized as follows. In Section 5.2, we formulate the machine-inventory model, and write down the generator matrix of the corresponding continuous time Markov chain. In Section 5.3, we construct a preconditioner for the generator matrix. In Section 5.4, we prove that the preconditioned linear system has singular values clustered around one. In Section 5.5, we give a cost analysis for our method and numerical examples are given to demonstrate the fast convergence rate of our method in Section 5.6. A summary is given in Section 5.7 to conclude this chapter.

5.2 The Machine-Inventory System

In this section, we construct the generator matrix for the machine-inventory system. Under an HPP policy, the maximum possible inventory level is h. The maximum backlog of the system is m. Therefore the total number of possible inventory levels is $n = m + h + 1$. In practice the value of n can easily go up to thousands. The number of repairing phases l and the maximum batch k are much less than n, i.e. $k, l \ll n$. When the size of an arrival batch is greater than the allowable backlog left, part of the orders will be accepted up to the maximum allowable backlog m and the whole batch of orders will be rejected if the maximum backlog level m is already attained.

Again we let $\alpha(t)$ be the state of the machine at time t given by

$$\alpha(t) = \begin{cases} 0, & \text{if the machine is up,} \\ i, & \text{if the machine is under repair in phase } i \ (i = 1, \ldots, l), \end{cases}$$

and $x(t)$ be the inventory level at time t which takes integer values in $[-m, h]$. We assume that the initial inventory level is zero without loss of generality. Then the machine-inventory process

$$\{(\alpha(t), x(t)), t \geq 0\}$$

is a continuous time Markov chain taking values in the state space

$$S = \{(\alpha, x) : \alpha = 0, \ldots, l, \ x = -m, \ldots, h\}.$$

Each time when visiting a state, the process stays there for a random period of time that has an exponential distribution and is independent of the past behavior of the process. Let us derive the generator for the machine-inventory process. The machine states follow an irreducible Markov chain. If we order the machine state space in ascending order of phases of repair with the last state being the up state, then we can obtain the following $(l + 1) \times (l + 1)$ generator M for the machine states:

$$M = \begin{pmatrix} \mu_1 & & & & -\lambda \\ -\mu_1 & \mu_2 & & & \\ & -\mu_2 & \ddots & & \\ & & \ddots & \mu_l & \\ 0 & & & -\mu_l & \lambda \end{pmatrix}. \tag{5.1}$$

The generator M has positive diagonal entries, non-positive off-diagonal entries and each column sum of M is zero. The inventory level follows a continuous time Markov chain. If we order the inventory state space in descending order, the generator matrices for the inventory levels when the machine is up and down are given by the following $(m + h + 1) \times (m + h + 1)$ matrices T_1 and T_2, respectively. Here

$$
T_1 = \begin{pmatrix}
a & & & & & & & 0 \\
-a_1 & a & & & & & & \\
\vdots & -a_1 & \ddots & & & & & \\
-a_{k-1} & \vdots & \ddots & \ddots & & & & \\
-a_k & \ddots & \vdots & \ddots & \ddots & & & \\
& \ddots & -a_{k-1} & \vdots & -a_1 & a & \\
0 & & -a_k & -\gamma_{k-1} & \cdots & -\gamma_1 & 0
\end{pmatrix}
\tag{5.2}
$$

and

$$
T_2 = \begin{pmatrix}
a & -d & & & & & \\
-a_1 & a+d & -d & & & & \\
\vdots & -a_1 & a+d & \ddots & & & \\
-a_{k-1} & \vdots & \ddots & \ddots & \ddots & & \\
-a_k & \ddots & \vdots & -a_1 & \ddots & \ddots & \\
& \ddots & -a_{k-1} & \vdots & -a_1 & a+d & -d \\
0 & & -a_k & -\gamma_{k-1} & \cdots & -\gamma_1 & d
\end{pmatrix},
\tag{5.3}
$$

where

$$
\begin{cases}
a_i = aq_i, \\
\gamma_i = a \sum_{j=i}^{k} q_j.
\end{cases}
$$

We remark that each column sum of T_1 and T_2 is zero. Therefore the generators for the combined machine-inventory process under the hedging point policy are given by the following $[(l+1)(m+h+1)] \times [(l+1)(m+h+1)]$ matrices A, where

$$
A = \begin{pmatrix}
T_1 + \mu_1 I & & & & -\lambda I \\
-\mu_1 I & T_1 + \mu_2 I & & & \\
& -\mu_2 I & \ddots & & \\
& & \ddots & T_1 + \mu_l I & \\
0 & & & -\mu_l I & T_2 + \lambda I
\end{pmatrix}.
\tag{5.4}
$$

We are interested in the steady state of the system, i.e.

$$
p(\alpha, x) = \lim_{t \to \infty} \text{Prob}(\alpha(t) = \alpha, x(t) = x).
\tag{5.5}
$$

Let

$$
p(i) = \sum_{k=0}^{l} p(k, i), \quad i = -m, -(m-1), \ldots, 0, \ldots, h
$$

be the marginal steady-state probability of the inventory levels of the manufacturing system. The average running cost can be written as the sum of inventory cost and backlog cost. Therefore the average running cost for the machine-inventory system under the hedging point policy is as follows:

$$C(h) = c_I \sum_{i=1}^{h} ip(i) - c_B \sum_{i=-m}^{-1} ip(i). \tag{5.6}$$

We note that the generator A is irreducible, has zero column sum, positive diagonal entries; and non-positive off-diagonal entries. The steady-state probability distribution $\mathbf{p}$ exists and is then equal to the normalized form of the positive null vector. From the state ordering of A and (5.5), we have

$$\mathbf{p}_{in+j} = p(i,j), \quad i = 0, 1, \ldots, l, \text{ and } j = 1, 2, \ldots, n. \tag{5.7}$$

We consider an equivalent linear system

$$G\mathbf{x} \equiv (A + \mathbf{e}_1 \mathbf{e}_1^t)\mathbf{x} = \mathbf{e}_1, \tag{5.8}$$

where $\mathbf{e}_1 = (1, 0, \ldots, 0)^t$ is the $(l+1)(m+h+1)$ unit vector. Similar to the proof discussed in previous chapters, we have the following lemma.

Lemma 5.2.1. *The matrix G is non-singular.*

Usually, by making use of the block structure of the generator matrix A, classical iterative methods such as block Gauss-Seidel are applied in solving the steady-state probability distribution. However, in general the classical iterative methods have slow convergence rate; see the numerical results in Section 5.6. In order to save computational cost, we employ Conjugate Gradient (CG) type methods to solve the steady-state probability distribution. To speed up the convergence, a preconditioner C is introduced.

5.3 Construction of Preconditioner

In this section, we construct a preconditioner from the near-Toeplitz structure of the diagonal blocks of A. The method of construction here is a little bit different from that in Chapter 2. We consider a circulant approximation of the matrices T_1 and T_2. Recall in Chapter 2 that any $n \times n$ circulant matrix C_n is characterized by its first column (or the first row) and can be diagonalized by the discrete Fourier matrix F_n, i.e. $C_n = F_n \Lambda_n F_n^*$, where F_n^* is the conjugate transpose of F_n and Λ_n is a diagonal matrix containing the eigenvalues of C_n. The matrix vector multiplication of the forms $F_n \mathbf{y}$ and $F_n^* \mathbf{y}$ can be obtained in $O(n \log n)$ operations by the Fast Fourier Transform (FFT). By completing T_1 and T_2 to circulant matrices, we define the circulant approximation $c(T_1)$ and $c(T_2)$ of T_1 and T_2 as follows:

$$c(T_1) = \begin{pmatrix} a & 0 & & -a_k & \cdots & -a_2 & -a_1 \\ -a_1 & a & \ddots & & -a_k & \vdots & -a_2 \\ \vdots & -a_1 & \ddots & \ddots & & \ddots & \vdots \\ -a_{k-1} & \vdots & \ddots & \ddots & \ddots & & -a_k \\ -a_k & \ddots & \vdots & \ddots & \ddots & \ddots & \\ & \ddots & \ddots & \vdots & \ddots & a & 0 \\ 0 & & -a_k & -a_{k-1} & \cdots & -a_1 & a \end{pmatrix} \qquad (5.9)$$

and

$$c(T_2) = \begin{pmatrix} a+d & -d & 0 & & -a_k & \cdots & -a_1 \\ -a_1 & a+d & -d & \ddots & & \ddots & \vdots \\ \vdots & -a_1 & \ddots & \ddots & \ddots & & -a_k \\ -a_k & \vdots & \ddots & \ddots & \ddots & \ddots & \\ 0 & -a_k & \vdots & \ddots & \ddots & \ddots & 0 \\ & \ddots & \ddots & \vdots & & -a_1 & a+d & -d \\ -d & & 0 & -a_k & \cdots & -a_1 & a+d \end{pmatrix}. \qquad (5.10)$$

From (5.9) and (5.10) we define the circulant approximation $c(A)$ of the generator matrix A as follows:

$$c(A) = \begin{pmatrix} c(T_1) + \mu_1 I & & & & -\lambda I \\ -\mu_1 I & c(T_1) + \mu_2 I & & & \\ & -\mu_2 I & \ddots & & \\ & & \ddots & c(T_1) + \mu_l I & \\ 0 & & & -\mu_l I & c(T_2) + \lambda I \end{pmatrix}. (5.11)$$

From (5.9) and (5.10) and Davis [62], we have the following lemmas.

Lemma 5.3.1. $\mathrm{Rank}(c(T_1) - T_1) = \mathrm{Rank}(c(T_2) - T_2) = k + 1$.

Lemma 5.3.2. *The matrices $c(T_1)$ and $c(T_2)$ can be diagonalized by the discrete Fourier transform F_n. The eigenvalues of $c(T_1)$ and $c(T_2)$ are given by*

$$F_n^* c(T_1) F_n = \mathrm{Diag}(\nu_1, \nu_2, \ldots, \nu_n)^t$$

and

$$F_n^* c(T_2) F_n = \mathrm{Diag}(\xi_1, \xi_2, \ldots, \xi_n)^t,$$

where

$$\begin{cases} \nu_j = \sum_{r=1}^{k} a_r(1 - e^{\frac{2\pi(r)(j)}{n}i}), & j = 1, 2, \ldots, n \\ \xi_j = d(1 - e^{\frac{2\pi(n-1)(j-1)}{n}i}) + \sum_{r=1}^{k} a_r(1 - e^{\frac{2\pi(r)(j)}{n}i}), & j = 1, 2, \ldots, n. \end{cases}$$

We note that the diagonal blocks of the $c(A)$ can be diagonalized by the discrete Fourier transform F_n. Moreover, there exists a permutation matrix P such that

$$P^t \cdot (I_{l+1} \otimes F_n^*) \cdot c(A) \cdot (I_{l+1} \otimes F_n) \cdot P = \text{Diag}(C_1, C_2, \ldots, C_n), \qquad (5.12)$$

where

$$C_i = \begin{pmatrix} \mu_1 + \nu_i & & & & -\lambda \\ -\mu_1 & \mu_2 + \nu_i & & & \\ & -\mu_2 & \ddots & & \\ & & \ddots & \mu_l + \nu_i & \\ 0 & & & -\mu_l & \lambda + \xi_i \end{pmatrix}. \qquad (5.13)$$

We note that all C_i except C_1 are strictly diagonal dominant and therefore they are non-singular. Moreover, we observe that $C_1 = M$ (cf. (5.1)). By similar argument in the proof of Lemma 5.2.1, we have $\tilde{C}_1 = (C_1 + \mathbf{e}_1 \mathbf{e}_1^t)$ is non-singular. Our preconditioner C is then defined as

$$C = (I_{l+1} \otimes F_n) \cdot P \cdot \text{Diag}(\tilde{C}_1, C_2, \ldots, C_n) \cdot P^t \cdot (I_{l+1} \otimes F_n^*). \qquad (5.14)$$

5.4 Convergence Analysis

In this section, we study the convergence rate of the PCG method when $n = m + h + 1$ is large. In practice, the number of possible inventory states n is much larger than the number of machine states l in the manufacturing systems and can easily go up to thousands. We prove that if all the parameters a, d, λ, l, k, q_i, and μ_i are fixed independent of n, then the preconditioned system GC^{-1} has singular values clustered around one as n tends to infinity. Hence when CG type methods are applied to solving the preconditioned system, we expect fast convergence. Numerical examples are given in Section 5.6 to illustrate our claim. We begin the proof with the following lemma.

Lemma 5.4.1. We have $\text{rank}(G - C) \leq (k+1)(l+1) + 2$.

Proof. We note that by (5.8), we have $\text{rank}(G - A) = 1$. From Lemma 5.3.1, we see that

$$\text{rank}(A - c(A)) = (k+1)(l+1).$$

From (5.13) and (5.14), $c(A)$ and C differ by a rank one matrix. Therefore, we have

$$
\begin{aligned}
\mathrm{rank}(G - C) &\leq \mathrm{rank}(G - A) + \mathrm{rank}(A - c(A)) + \mathrm{rank}(c(A) - C) \\
&= 1 + (k + 1)(l + 1) + 1 \\
&= (k + 1)(l + 1) + 2.
\end{aligned}
$$

$\square$

Proposition 5.4.1. *The preconditioned matrix GC^{-1} has at most $2((k + 1)(l + 1) + 2)$ singular values not equal to one.*

Proof. We first note that

$$
GC^{-1} = I + (G - C)C^{-1} \equiv I + L_1,
$$

where

$$
\mathrm{rank}(L_1) \leq (k + 1)(l + 1) + 2
$$

by Lemma 5.4.1. Therefore

$$
C^{-*}G^*GC^{-1} - I = L_1^*(I + L_1) + L_1,
$$

is a matrix of rank at most $2((k + 1)(l + 1) + 2)$. $\square$

Thus the number of singular values of GC^{-1} that are different from one is a constant independent of n. Recall that in Chapter 3, in order to show the fast convergence of preconditioned conjugate gradient type methods with our preconditioner C, one still needs to estimate $\sigma_{\min}(GC^{-1})$, the smallest singular value of GC^{-1}. If $\sigma_{\min}(GC^{-1})$ is uniformly bounded away from zero independent of n, then the method converges in $O(1)$ iterations; and if $\sigma_{\min}(GC^{-1})$ decreases like $O(n^{-\alpha})$ for some $\alpha > 0$, then the method converges in at most $O(\log n)$ steps.

In the following, we are going to show that even in the worst case where $\sigma_{\min}(GC^{-1})$ decreases in an order faster than $O(n^{-\alpha})$ for any $\alpha > 0$ (e.g. like $O(e^{-n})$), we can still have a fast convergence rate. The proof is similar to that in Section 3.5. Note that in this case the preconditioned system is very ill-conditioned. Again our trick is to consider a regularized equation as follows:

$$
C^{-*}(G^*G + n^{-4-\beta}I)C^{-1}\mathbf{w} = C^{-*}G^*\mathbf{e}_1 \tag{5.15}
$$

where β is any positive constant. In the following, we prove that the regularized preconditioned matrix

$$
C^{-*}(G^*G + n^{-4-\beta}I)C^{-1}
$$

has eigenvalues clustered around one and its smallest eigenvalue decreases at a rate no faster than $O(n^{-4-\beta})$. Hence PCG type methods will converge in at most $O(\log n)$ steps when applied to solving the preconditioned linear system (5.15). Moreover, we prove that the 2-norm of the error introduced by the regularization tends to zero at a rate of $O(n^{-\beta})$. In order to prove our claim, we have to get an estimate of the upper and lower bounds for $||C^{-1}||_2$.

Lemma 5.4.2. *Let the parameters a, d, λ, l, k, q_i, and μ_i be independent of n. Then there exist positive constants τ_1 and τ_2 independent of n such that $\tau_1 \leq ||C^{-1}||_2 \leq \tau_2 n^2$.*

Proof. We first prove the left hand side inequality. From (5.14), we see that C is unitarily similar to a diagonal block matrix. We therefore have

$$||C||_2 = \max \left\{ ||\tilde{C}_1||_2, ||C_2||_2, \ldots, ||C_n||_2 \right\}.$$

From (5.13), it is obvious that $||C_i||_1$, $||C_i||_\infty$, $||\tilde{C}_1||_1$, and $||\tilde{C}_1||_\infty$ are all bounded above by

$$\frac{1}{\tau_1} = 2(\max_i \{\mu_i, \lambda\} + a + d), \quad 1 \leq i \leq n.$$

Using the inequality

$$|| \cdot ||_2 \leq \sqrt{|| \cdot ||_1 || \cdot ||_\infty},$$

we see that $||\tilde{C}_1||_2$ and $||C_i||_2$, $(i = 2, \ldots, n)$, are all bounded above by $1/\tau_1$. Thus $||C||_2 \leq 1/\tau_1$ and hence $\tau_1 \leq ||C^{-1}||_2$.

Next we prove the right hand side inequality. We note by (5.14) that

$$||C^{-1}||_2 = \max \left\{ ||\tilde{C}_1^{-1}||_2, ||C_2^{-1}||_2, \ldots, ||C_n^{-1}||_2 \right\}.$$

and $||\tilde{C}_1^{-1}||_2$ is bounded independent of n.

To estimate $||C_i^{-1}||_2$, $(i = 2, \ldots, n)$, we consider the matrix (cf. (5.13))

$$W_i = C_i \cdot \text{Diag} \left(\frac{1}{\mu_1}, \cdots, \frac{1}{\mu_l}, \frac{1}{\lambda} \right) \equiv C_i U = W_{i1} + W_2. \tag{5.16}$$

From (5.13) W_{i1} is the diagonal matrix

$$\text{Diag} \left(\frac{\nu_i}{\mu_1}, \ldots, \frac{\nu_i}{\mu_l}, \frac{\xi_i}{\lambda} \right) \tag{5.17}$$

and W_2 is the $(l+1) \times (l+1)$ circulant matrix with first row equals to

$$(1, 0, \ldots, 0, -1).$$

The eigenvalues of W_2 are given by

$$(1 - e^{\frac{2\pi j}{l+1} i}), \quad j = 0, 1, 2, \ldots, l \tag{5.18}$$

and both W_2 and W_2^* can be diagonalized by the discrete Fourier transform F_{l+1}. Thus the eigenvalues of the real symmetry matrix $(W_2 + W_2^*)$ are

$$2\sin^2\left(\frac{\pi j}{l+1}\right), \quad j = 0, 1, \ldots, l. \tag{5.19}$$

Thus $(W_2 + W_2^*)$ is a symmetric positive semi-definite matrix. From Lemma 3.5.1, we observe that

$$\begin{aligned}
\mathrm{Re}(\nu_j) &= \sum_{r=1}^{k} a_r \left(1 - \cos(\frac{2\pi r j}{n})\right) \\
&= 2 \sum_{r=1}^{k} a_r \sin^2\left(\frac{\pi r j}{n}\right) \\
&\geq 2 \sum_{r=1}^{k} a_r \sin^2\left(\frac{\pi}{n}\right) \\
&= 2a \sin^2\left(\frac{\pi}{n}\right).
\end{aligned}$$

Similarly, we have

$$\mathrm{Re}(\xi_j) \geq 2\left(a \sin^2\left(\frac{\pi}{n}\right) + d\sin^2\left(\frac{\pi}{n}\right)\right). \tag{5.20}$$

Here $\mathrm{Re}(z)$ is the real part of the complex number z. By the inequality

$$\sin(\theta) \geq \min\left\{\frac{2\theta}{\pi}, 2\left(1 - \frac{\theta}{\pi}\right)\right\} \quad \forall \theta \in [0, \pi], \tag{5.21}$$

we have $\mathrm{Re}(\nu_j) \geq 2a/n^2$ and $\mathrm{Re}(\xi_j) \geq 2(a + d)/n^2$. From (5.17),(5.16) and the inequality $|z| \geq |\mathrm{Re}(z)|$, we get

$$\lambda_{\min}(W_{i1} + W_{i1}^*) \geq \frac{2a}{n^2 \max\{\mu_i, \lambda\}}. \tag{5.22}$$

By Weyl's theorem (see Chapter 3 or Horn and Johnson [76, p. 181]), we have

$$\lambda_{\min}(W_i + W_i^*) = \lambda_{\min}((W_{i1} + W_{i1}^*) + (W_2 + W_2^*))$$

$$\geq \frac{2a}{n^2 \max\{\mu_i, \lambda\}}, \quad i = 2, 3, \ldots, n.$$

Thus by Lemma 3.5.1 we have

$$\|W_i^{-1}\|_2 \leq \frac{n^2 \max\{\mu_i, \lambda\}}{a}$$

and therefore by (5.16)

$$\|C_i^{-1}\|_2 = \|UU^{-1}C_i^{-1}\|_2$$
$$\leq \|U\|_2\|(C_iU)^{-1}\|_2 = \|U\|_2\|W_i^{-1}\|_2$$
$$\leq a^{-1}n^2 \max\{\mu_i, \lambda\}\|U\|_2,$$

where $\|U\|_2$ is independent of n. Finally, we get

$$\|C^{-1}\|_2 \leq \max\{\|\tilde{C}_1^{-1}\|_2, \frac{\max\{\mu_i, \lambda\}\|U\|_2}{a}\}n^2 \equiv \tau_2 n^2,$$

where τ_2 is independent of n.   □

From Lemma 5.4.2, we remark that our preconditioner C will never be ill-conditioned even if G is ill-conditioned. We have also the following proposition.

Proposition 5.4.2. *Let the parameters $a, d, \lambda, k, q_i,$ and μ_i be independent of n. Then for any positive β, the regularized preconditioned matrix*

$$C^{-*}(G^*G + n^{-4-\beta}I)C^{-1} \tag{5.23}$$

has eigenvalues clustered around one and the smallest eigenvalue decreases at a rate no faster than $O(n^{-4-\beta})$. Furthermore, the error introduced by the regularization is of the order $O(n^{-\beta})$.

Thus we conclude that PCG type methods applied to (5.15) with $\beta > 0$ will converge in at most $O(\log n)$ steps; see Chan [19, Lemma 3.8.1]. To minimize the error introduced by the regularization, one can choose a large β. Recall that regularization is required only when the smallest singular value of the matrix GC^{-1} tends to zero faster than $O(n^{-\alpha})$ for any $\alpha > 0$.

In general we do not know if G is ill-conditioned, i.e. $\sigma_{min}(G^*G)$ decreases like $O(e^{-n})$. However, we may always assume this is the case and apply CG method to the regularized equation (5.15). From Proposition 5.4.2, the error in 2-norm due to the regularization (cf. Lemma 5.4.2) is given by

$$\|C^{-*}C^{-1}\|_2 n^{-4-\beta} \leq \|C^{-*}\|_2\|C^{-1}\|_2 n^{-4-\beta} \leq \tau_2^2 n^{-\beta}. \tag{5.24}$$

Recall that (cf. Lemma 5.4.2)

$$\tau_2 = \max\left\{\|\tilde{C}_1^{-1}\|_2, \frac{\max\{\mu_i, \lambda\}\|U\|_2}{a}\right\}, \tag{5.25}$$

then the error can be reduced as small as possible by choosing a large β. We note, however, that in all our numerical tests in Section 5.6, we found that there is no need to add the regularization.

5.5 Computational and Memory Cost Analysis

From (5.9),(5.10),(5.11) and (5.13), the construction of our preconditioner C need no cost. The main computational cost of our method comes from the matrix vector multiplication of the form $G\mathbf{x}$, and solving the preconditioner system $C\mathbf{y} = \mathbf{r}$. By making use of the band structure of G, the matrix vector multiplication $G\mathbf{x}$ can be done in $O(k(l+1)n)$ operations. The solution for $C\mathbf{y} = \mathbf{r}$ can be written as follows (cf. 5.14):

$$\mathbf{y} = (I_{l+1} \otimes F_n) \cdot P \cdot \mathrm{Diag}(\tilde{C}_1^{-1}, C_2^{-1}, \dots, C_n^{-1}) \cdot P^t \cdot (I_{l+1} \otimes F_n^*)\mathbf{r}. \quad (5.26)$$

The matrix vector multiplication of the forms $F_n\mathbf{x}$ and $F_n^*\mathbf{x}$ can be done in $O(n \log n)$ operations. By using the following lemma, the solution of the linear system

$$\mathrm{Diag}(\tilde{C}_1^{-1}, C_2^{-1}, \dots, C_n^{-1})\mathbf{y} = \mathbf{b}$$

can be obtained in $O((l+1)n)$ operations. Hence the cost for solving (5.26) is $O((l+1)n \log n + (l+1)n)$.

Lemma 5.5.1. *(Sherman-Morrison-Woodbury formula.) Let Z be an invertible $n \times n$ matrix and U and V are $n \times k$ matrices. Suppose that $I + V^t Z^{-1} U$ is invertible then*

$$(Z + UV^t)^{-1} = Z^{-1} - Z^{-1}U(I + V^t Z^{-1} U)^{-1} V^t Z^{-1}.$$

Proof. See Exercise 5.2 or Golub [70, p. 51]. $\qquad\Box$

Proposition 5.5.1. *The linear system $H\mathbf{x} = \mathbf{b}$ can be solved in $O(l)$ operations if $\det(H) \neq 0$, where*

$$H = \begin{pmatrix} e_1 & & & & -c \\ -b_1 & e_2 & & & \\ & -b_2 & \ddots & & \\ & & & \ddots & e_l \\ 0 & & & & -b_l & e_{l+1} \end{pmatrix}. \quad (5.27)$$

Proof. Let L be the lower triangular part (including the main diagonal) of the matrix H. We have

$$H = L + (1, 0, \dots, 0)^t(0, \dots, 0, -c) \equiv L + \mathbf{u}\mathbf{v}^t.$$

By using the Sherman-Morrison-Woodbury formula, the inverse of H can be expressed as follows:

$$H^{-1} = L^{-1}(I - \mathbf{u}(1 + \mathbf{v}^t L^{-1} \mathbf{u})^{-1} \mathbf{v}^t L^{-1}), \quad (5.28)$$

provided that

$$\Delta = 1 + \mathbf{v}^t H^{-1} \mathbf{u} \neq 0.$$

By direct verification,

$$\Delta = 1 - \frac{b_1 b_2 \cdots b_l c}{e_1 e_2 \cdots e_{l+1}}, \quad \text{and} \quad \det(H) = e_1 e_2 \cdots e_{l+1} - b_1 b_2 \cdots b_l c.$$

Therefore $\Delta \neq 0$ if and only if $\det(H) \neq 0$. Hence we may solve the linear system $H\mathbf{x} = \mathbf{b}$ by using (5.28). It is straightforward to check that we need only $O(l)$ operations to obtain the solution. $\qquad\square$

We conclude that in each iteration of the PCG method, we need $O((l + 1)n \log n)$ operations. The cost per iteration of the Block Gauss-Seidel (BGS) method is of $O(k(l + 1)n)$. This can be done by making use of the band structure of the diagonal blocks of the generator matrix A. Although the PCG method requires an extra $O(\log n)$ operations in each iteration, the fast convergence rate of our method can more than compensate for this minor overhead (see the numerical examples below). In Propositions 5.4.1 and 5.4.2, we have proved the preconditioned linear system and the preconditioned linear system with regularization (5.23) has singular values clustered around one, so the total number of operations for solving the steady-state probability vector is at most $O((l + 1)n \log^2 n)$. Both PCG and BGS require $O((l + 1)n)$ memory. Clearly at least $O((l + 1)n)$ memory is required to store the approximated solution in each iteration.

5.6 Numerical Examples

In this section, we employ a CGS method to solve the preconditioned system. The method does not require the transpose of the iteration matrix $G^{-1}C$. We compare our PCG method with the classical iterative method, the Block Gauss-Seidel (BGS) in the following numerical examples. In our examples, we assume that the machine has a mean up time of three weeks on average, i.e. $\lambda = 1/3$. If the machine is down, it takes one week for the l phases of repairing and the mean time of repairing in each phase is assumed to be equal, i.e.

$$\frac{1}{\mu_1} = \cdots = \frac{1}{\mu_l} = \frac{1}{l}.$$

Usually, the number of repairing phases l is smaller than 8, so we test our method for l up to 8. Here we assume that the mean time for arrival of demand is one week and the mean time for the machine to produce one unit of product is one day. Therefore we have $a = 1$ and $d = 7$. The distribution for the batch size is set as follows:

$$q_1 = 1/2, \ q_2 = 1/4, \ q_3 = 1/8, \ q_4 = 1/8, \quad \text{and} \quad q_i = 0 \quad \text{for} \quad i \geq 5.$$

The stopping criterion for both PCGS and BGS is

$$||A\mathbf{p}_k||_2 < 10^{-10},$$

where $\mathbf{p}_k$ is the approximated solution obtained at the kth iteration. The initial guess for both methods is the unit vector $\mathbf{e}_1 = (1, 0, \ldots, 0)^t$. We give the number of iterations for convergence of PCGS (Tables 5.1) for different values of l. The symbols I, C, BGS represent the methods used, namely, CGS without preconditioner, CGS with preconditioner C, and the BGS method. We see that the number of iterations for convergence is roughly constant independent of $n = m + h + 1$. The symbol "**" signifies that the number of iterations is greater than 200.

Table 5.1. Number of iterations for convergence

n	$l = 1$			$l = 2$			$l = 4$			$l = 8$		
	I	C	BGS	I	C	BGS	I	C	BGS	I	C	BGS
16	41	7	34	53	10	34	74	14	34	47	24	34
64	**	7	**	**	10	**	**	13	**	**	22	**
256	**	8	**	**	10	**	**	13	**	**	21	**
1024	**	8	**	**	10	**	**	13	**	**	21	**

Next we consider the optimal hedging point h^* of the example; we set $l = 2$, $m = 50$, and the maximum inventory level to be 200. Moreover, the inventory cost c_I and backlog cost c_B per unit of product are 50 and 2000, respectively. We keep the values of a, d, μ_i, and q_i in the above examples. We note that the average production rate of the machine and the mean arrival rate of demand per week are $\frac{3}{4}d$ and $\frac{15}{8}$ respectively. To meet the average demand rate, we should have $d > \frac{5}{2}$. In the following table (Table 5.2), we test for different values of $d > \frac{5}{2}$. We give the optimal values of hedging point h^* and its corresponding average running cost per week.

Table 5.2. The optimal h^* for different d

d	6	7	8	9	10
h^*	59	46	39	35	32
$C(h^*)$	3077	2430	2090	1882	1743

5.7 Summary and Discussion

An unreliable one-machine manufacturing system has been studied in this chapter. Two special features of the system are the batch arrivals of demand and the Erlangian repairing process.

One may extend our method of circulant approximation to a manufacturing system of q unreliable machines in parallel under Erlangian repairing process and batch arrivals. Let us construct the generator matrix for this machine-inventory system. Under the hedging point policy, the maximum possible inventory level is h. The maximum backlog of the system is m. Therefore the total number of possible inventory levels is $n = (m + h + 1)$. For each machine j, $(j = 1, \ldots, q)$, let Q_j be the generator matrix of the machine states and Λ_j be the corresponding production rate matrix. Here

$$
Q_j = \begin{pmatrix}
\mu_{j1} & & & & -\lambda_j \\
-\mu_{j1} & \mu_{j2} & & & \\
& -\mu_{j2} & \ddots & & \\
& & \ddots & \mu_{jl} & \\
0 & & & -\mu_{jl} & \lambda_j
\end{pmatrix}
$$

and

$$
\Lambda_j = \begin{pmatrix}
0 & & & & \\
& 0 & & & \\
& & \ddots & & \\
& & & 0 & \\
& & & & d_j
\end{pmatrix}.
$$

Since there are q machines, there are l^q states for the system of machines. We denote the set of machines states by Ω. The superposition of the q machines forms another MMPP and is characterized by the following $l^q \times l^q$ generator matrix:

$$
Q = (Q_1 \otimes I_l \otimes \cdots \otimes I_l) + (I_l \otimes Q_2 \otimes I_l \otimes \cdots \otimes I_l) + \cdots + (I_l \otimes \cdots \otimes I_l \otimes Q_q).
$$

The corresponding production rate matrix is given by

$$
\Lambda = (\Lambda_1 \otimes I_l \otimes \cdots \otimes I_l) + (I_l \otimes \Lambda_2 \otimes I_l \otimes \cdots \otimes I_l) + \cdots + (I_l \otimes \cdots \otimes I_l \otimes \Lambda_q).
$$

We let $\alpha(t)$ be the state of the system of machines at time t. Therefore $\alpha(t)$ has l^q possible states. The inventory level takes integer value in $[-m, h]$, because we allow maximum backlog of m and the hedging point is h. We let $x(t)$ be the inventory level at time t. The machine inventory process $\{(\alpha(t), x(t)), t \geq 0\}$ forms an irreducible continuous time Markov chain in the state space

$$
\{(\alpha, x) \mid \alpha \in \Omega, \ x = h, h - 1, \ldots, 0, \ldots, -m\}.
$$

If we order the state space of the machine inventory process lexicographically, we get the following $(nl^q) \times (nl^q)$ $(n = m + h + 1)$ generator matrix B for the machine-inventory system:

$$B = I_n \otimes Q + R \otimes \Lambda + T_1 \otimes I_{l^q}$$

where I_n and I_{l^q} are the $n \times n$ identity matrix and the $l^q \times l^q$ identity matrix respectively, T_1 (cf. 5.2) and

$$R = \begin{pmatrix} 0 & -1 & & 0 \\ & 1 & \ddots & \\ & & \ddots & \\ & & 1 & -1 \\ 0 & & & 1 \end{pmatrix}.$$

One may construct a preconditioner for B by taking the circulant approximation as follows:

$$c(B) = I_n \otimes Q + c(R) \otimes \Lambda + c(T_1) \otimes I_{l^q} \tag{5.29}$$

where

$$c(R) = \begin{pmatrix} 1 & -1 & & 0 \\ & 1 & \ddots & \\ & & \ddots & \\ & & 1 & -1 \\ -1 & & & 1 \end{pmatrix}.$$

One can show that $c(B)$ is singular and has a null space of dimension one. Furthermore Rank$(B - c(B))$ is independent of n and $c(B)$ is similar to a diagonal block matrix. Thus by making a rank one perturbation, one can obtain a good preconditioner for B.

Exercises

5.1 Prove Proposition (5.4.2).

5.2 Show that for any two invertible matrices A and B we have

$$B^{-1} = A^{-1} - B^{-1}(B - A)A^{-1}.$$

Hence prove the Sherman-Morrison-Woodbury formula:

$$(Z + UV^t)^{-1} = Z^{-1} - Z^{-1}U(I + V^t Z^{-1}U)^{-1}V^t Z^{-1}.$$

5.3 Show that the steady-state probability distribution for the generator matrix M in (5.1) is given by

$$\alpha^{-1} \left(\frac{1}{\mu_1}, \ldots, \frac{1}{\mu_l}, \frac{1}{\lambda} \right)^t$$

where

$$\alpha = \frac{1}{\mu_1} + \cdots + \frac{1}{\mu_l} + \frac{1}{\lambda}.$$

5.4 Show that $c(B)$ in (5.29) has a null space of dimension one. Show also that $\mathrm{Rank}(B - c(B))$ is independent of the size of inventory levels n and $c(B)$ is similar to a diagonal block matrix.

6. Flexible Manufacturing Systems of Unreliable Machines

6.1 Introduction

In this chapter we study Markovian models for the performance evaluation of *Flexible Manufacturing Systems* (FMSs). In recent years there has been an increasing role of computers in manufacturing. One important area of *Computer Aided Manufacturing* (CAM) is FMS. The advantage of an FMS is that it can reduce the in the system and increases the machine utilization when suitable production policy is implemented. Moreover, it can also reduce manufacturing lead time and labor cost. However, the setting up cost and the maintenance cost of machines are high in an FMS; see Buzacott and Shanthikumar [17] for instance. Due to the high capital investment, an FMS is only considered to operate economically if there is a high level of system performance. The system design and management are therefore important front considerations for setting up an FMS. Mathematical modeling can help with decisions required to design and manage an FMS. Queuing theory is a useful tool for modeling manufacturing systems; see for instance [44, 46, 110]. In fact, most analytical models describe an FMS as a queuing system, in which the customers are jobs to be processed or product in inventory and the servers are simply the reliable machines (workstations) in the system; see for instance Buzacott and Yao [18]. However, the assumption that the machines are reliable can greatly affect the performance evaluation of an FMS and therefore should be taken into account in any proposed model.

In this chapter, we consider FMSs of multiple unreliable machines producing one type of product. When a machine breaks down, it is subject to a repairing process if there is maintenance facility available, otherwise it will queue up and wait for repairing. The repairing process is based on a first-come-first-served principle. There are r ($\leq s$) maintenance facilities in the FMS where s is the number of machines in the system. For simplicity of discussion, we assume the machines are all identical. The model and methodology proposed here are still applied when the machines have different processing rates and failure rates. We assume that the normal time and repairing time of each machine are exponentially distributed.

Usually, a proper positive inventory level is maintained to hedge against the uncertainties in demand or breakdowns of the machine. We employ the HPP policy discussed in Chapter 4 as the production control. In the model, we assume that the inter-arrival time for a demand and the processing time for one unit of product are exponentially distributed. The demand is served in first-come-first-in-served principle. Furthermore, we allow a maximum backlog of m in the system. It turns out that the combined machine-inventory process is an irreducible continuous time Markov chain. We are interested in solving the steady-state probability distribution for the FMS because many important performance measures such as the average system running cost, system throughput, and machine utilization rate can be written in terms of the probability distribution. Moreover, the average profit for the system can also be written in terms of the probability distribution. Therefore the optimal hedging point can be obtained by varying different possible values of h. The following notation will be used for our discussion throughout the chapter.

(i) λ^{-1}, the mean waiting time for a demand,
(ii) μ^{-1}, the mean processing time for one unit of product,
(iii) σ^{-1}, the mean repair time for a machine,
(iv) γ^{-1}, the mean up time of a machine,
(v) s, number of machines (workstations) in the system,
(vi) r, number of maintenance facilities,
(vii) m, maximum allowable backlog,
(viii) b, the maximum inventory capacity,
(ix) $h(\leq b)$, the hedging point,
(x) $c_I > 0$, the unit inventory cost,
(xi) $c_B > 0$, the unit backlog cost,
(xii) $c_O > 0$, operation cost of a machine per unit time,
(xiii) $c_R > 0$, repairing cost of a machine per unit time,
(xiv) $c_P > 0$, profit of a product.

The remainder of this chapter is organized as follows. In Section 6.2, we formulate the FMS as a machine-inventory model, and give the generator matrix of system. In Section 6.3, the conjugate gradient method is applied to solve the system steady-state probability distribution. We construct a preconditioner for the system by taking a circulant approximation of the generator matrix. A convergence rate proof is also given in Section 6.4. In Section 6.5, we give a cost analysis of the proposed numerical algorithm. In Section 6.6, numerical examples are given to demonstrate the fast convergence rate of the proposed method. Applications of our proposed model are also illustrated in Section 6.7. Finally a summary is given in Section 6.8 to conclude the chapter.

6.2 The Machine-Inventory Model for FMS

In this section, we construct the generator matrix for the machine-inventory system. Under a HPP policy, the maximum possible inventory level is h. Since the maximum allowable backlog is m, the total number of possible inventory levels is $n = m+h+1$. In practice the value of n can easily go up to thousands. The number of machines which are normal can take values in $\{0, 1, \ldots, s\}$. We let $\alpha(t)$ be the number of machines which are normal at time t and $x(t)$ be the inventory level at time t. Then the machine inventory process

$$\{(\alpha(t), x(t)), t \geq 0\}$$

is a continuous time Markov chain taking values in the state space

$$S = \{(\alpha(t), x(t)) : \alpha(t) = 0, \ldots, s, \ x(t) = -m, \ldots, h\}.$$

Let us derive the generator for the machine-inventory process. If we order the machine states and the inventory states lexicographically: the machine states are in ascending order of number of normal machines and the inventory states are in descending order. Then the generator A is given by

$$A = \begin{pmatrix} N_0 & -\gamma & & & & & 0 \\ -r\sigma & N_1 & -2\gamma & & & & \\ & \ddots & \ddots & \ddots & & & \\ & & -r\sigma & N_{(s-r+1)} & -(s-r+2)\gamma & & \\ & & & \ddots & \ddots & \ddots & \\ & & & & -2\sigma & N_{(s-1)} & -s\gamma \\ 0 & & & & & -\sigma & N_s \end{pmatrix}$$

and for $i = 0, 1, \ldots, s$,

$$N_i = \begin{pmatrix} \lambda & -i\mu & & & 0 \\ -\lambda & \lambda+i\mu & -i\mu & & \\ & \ddots & \ddots & \ddots & \\ & & -\lambda & \lambda+i\mu & -i\mu \\ 0 & & & -\lambda & i\mu \end{pmatrix} + \min\{s-i,r\}I_n \qquad (6.1)$$

$$\equiv M_i + \min\{s-i,r\}I_n.$$

Here I_n is the $n \times n$ identity matrix. In fact, M_i is the generator matrix for the inventory levels when there are i normal machines in the system.

We are interested in the steady-state probability distribution of the system, i.e.

$$p(\alpha, x) = \lim_{t \to \infty} \text{Prob}(\alpha(t) = \alpha, x(t) = x). \qquad (6.2)$$

This probability distribution is useful because we can derive the following system quantities. Let

$$p(i) = \sum_{k=0}^{s} p(k, i), \quad i = -m, -(m-1), \ldots, 0, \ldots, h \qquad (6.3)$$

be the marginal steady-state probability of the inventory levels of the FMS and

$$q(k) = \sum_{i=-m}^{h} p(k,i), \quad k = 0, 1, \ldots, s \qquad (6.4)$$

be the marginal steady-state probability of the machine states of the manufacturing system. The throughput of the system is given by

$$T(h,m) = \lambda(1 - p(-m)). \qquad (6.5)$$

The inventory cost can be written as the sum of inventory cost and backlog cost as follows:

$$I(h,m) = c_I \sum_{i=1}^{h} ip(i) - c_B \sum_{i=-m}^{-1} ip(i). \qquad (6.6)$$

The machine operating cost consists of repairing cost and running cost and is given by

$$M(h,m) = c_O \sum_{k=0}^{s} kq(k) + c_R \sum_{k=0}^{s-r} rq(k)$$
$$+ c_R \sum_{k=s-r+1}^{s} (s-k)q(k). \qquad (6.7)$$

Thus the average running cost and average profit are given by

$$C(h,m) = I(h,m) + M(h,m) \qquad (6.8)$$

and

$$P(h,m) = c_P T(h,m) - C(h,m) \qquad (6.9)$$

respectively.

We note that the generator matrix A is irreducible, has zero column sum, positive diagonal entries, and non-positive off-diagonal entries. The steady-state probability distribution $\mathbf{p}$ exists and is then equal to the normalized form of the positive null vector. From the state ordering of A and (6.2), we have

$$\mathbf{p}_{in+j} = p(i,j), \quad i = 0, 1, \ldots, s, \text{ and } j = 1, 2, \ldots, n. \qquad (6.10)$$

Again we consider the equivalent linear system

$$G\mathbf{x} \equiv (A + \mathbf{e}_1 \mathbf{e}_1^t)\mathbf{x} = \mathbf{e}_1, \qquad (6.11)$$

where $\mathbf{e}_1 = (1, 0, \ldots, 0, 0)^t$ is the $(s+1)(m+h+1)$ unit vector. The steady-state probability distribution vector is the normalized solution of (6.11). It is clear that

Lemma 6.2.1. *The matrix G is non-singular.*

Classical iterative methods have slow convergence rates in general; see the numerical results in Section 6.6. We employ Conjugate Gradient (CG) type methods to solve the steady-state probability distribution. To speed up the convergence, a preconditioner C is introduced. In the next section the PCG method is applied to solving the preconditioned linear system with a circulant-based preconditioner. We first construct circulant preconditioners for the preconditioned linear system. A convergence rate proof is given when the states of the inventory level are large.

6.3 Construction of Preconditioner

In this section, we construct preconditioners by taking a circulant approximation of the system generator matrix A. We note that in fact, we can write

$$A = \text{Diag}(M_0, M_1, \ldots, M_s) + T \otimes I_n \qquad (6.12)$$

where

$$T = \begin{pmatrix} r\sigma & -\gamma & & & & 0 \\ -r\sigma & \gamma + r\sigma & -2\gamma & & & \\ & \ddots & \ddots & \ddots & & \\ & & -r\sigma & -(s-r+2)\gamma & & \\ & & & \ddots & \ddots & \ddots \\ & & & & -2\sigma & (s-1)\gamma+\sigma & -s\gamma \\ 0 & & & & & -\sigma & s\gamma \end{pmatrix} \qquad (6.13)$$

Here T is the generator matrix for the machine states. We consider a circulant approximation of the matrices $M_0, M_1, \ldots, M_s$. It has been shown in Chapter 2 that any $n \times n$ circulant matrix S_n is characterized by its first column (or the first row) and can be diagonalized by the discrete Fourier matrix F_n, i.e. $S_n = F_n \Lambda_n F_n^*$, where F_n^* is the conjugate transpose of F_n and Λ_n is a diagonal matrix containing the eigenvalues of S_n. The matrix vector multiplication of the forms $F_n y$ and $F_n^* y$ can be obtained in $O(n \log n)$ operations by the FFT. By completing $M_0, M_1, \ldots, M_s$ to circulant matrices, the circulant approximations

$$c(M_0), c(T_1), \ldots, c(M_s)$$

are defined to be:

$$c(M_i) = \begin{pmatrix} \lambda + i\mu & -i\mu & & & -\lambda \\ -\lambda & \lambda + i\mu & -i\mu & & \\ & \ddots & \ddots & \ddots & \\ & & -\lambda & \lambda + i\mu & -i\mu \\ -i\mu & & & -\lambda & \lambda + i\mu \end{pmatrix}. \qquad (6.14)$$

We define the circulant approximation of A as

$$c(A) = \text{Diag}(c(M_0), c(M_1), \ldots, c(M_s)) + T \otimes I_n \qquad (6.15)$$

From (6.1) and (6.14) and Davis [62], we have the following lemmas.

Lemma 6.3.1. $\text{Rank}(c(M_i) - M_i) = 2$ and $\text{Rank}(c(A) - A) = 2(s + 1)$.

Lemma 6.3.2. The matrices $c(M_i)$ can be diagonalized by the discrete Fourier transform F_n. The eigenvalues of $c(M_i)$

$$F_n^* c(M_i) F_n = \text{Diag}(\xi_{i1}, \xi_{i2}, \ldots, \xi_{in})^t,$$

where

$$\xi_{ij} = i\mu(1 - e^{\frac{2\pi(n-1)(j-1)}{n}i}) + \lambda(1 - e^{\frac{2\pi(r)(j)}{n}i}), \quad j = 1, 2, \ldots, n.$$

We recall that the diagonal blocks of the $c(A)$ can be diagonalized by the discrete Fourier transform F_n. Moreover, there exists a permutation matrix P (see Chapter 5) such that

$$P^t \cdot (I_{l+1} \otimes F_n^*) \cdot c(A) \cdot (I_{l+1} \otimes F_n) \cdot P = \text{Diag}(C_1, C_2, \ldots, C_n), \qquad (6.16)$$

where

$$C_j = T + \text{Diag}(\xi_{0j}, \xi_{1j}, \ldots, \xi_{sj}), \quad j = 1, 2, \ldots, n. \qquad (6.17)$$

All C_j except C_1 are strictly diagonal dominant and therefore they are non-singular. In fact, $C_1 = T$. By similar argument in the proof of Lemma 6.2.1, we have $\tilde{C}_1 = (C_1 + \mathbf{f}\mathbf{f}^t)$ is non-singular. Here $\mathbf{f} = (1, 0, \ldots, 0)^t$. Our preconditioner C is then defined as

$$C = (I_{s+1} \otimes F_n) \cdot P \cdot \text{Diag}(\tilde{C}_1, C_2, \ldots, C_n) \cdot P^t \cdot (I_{s+1} \otimes F_n^*). \qquad (6.18)$$

6.4 Convergence Analysis

In this section, we study the convergence rate of the PCG method when $n = m + h + 1$ is large. In practice, the number of possible inventory states n is much larger than the number of machine states l in the manufacturing systems and can easily go up to thousands. We prove that if all the parameters $\lambda, \mu, \sigma, \gamma, s$, and r are fixed independent of n, then the preconditioned system GC^{-1} has singular values clustered around one as n tends to infinity. Hence when CG type methods are applied to solving the preconditioned system, we expect fast convergence. Numerical examples are given in Section 6.6 to verify our claim. We begin the proof with the following lemma.

Lemma 6.4.1. We have $\text{rank}(G - C) \leq 2s + 4$.

Proof. We note that by (6.11), we have $\text{rank}(G - A) = 1$. From Lemma 6.3.1, we see that

$$\text{rank}(A - c(A)) = 2(s + 1).$$

From (6.17) and (6.18), $c(A)$ and C differ by a rank one matrix. Therefore, we have

$$\begin{aligned}
\text{rank}(G - C) &\leq \text{rank}(G - A) + \text{rank}(A - c(A)) + \text{rank}(c(A) - C) \\
&= 1 + 2(s + 1) + 1 \\
&= 2s + 4.
\end{aligned}$$

$\square$

Proposition 6.4.1. *The preconditioned matrix GC^{-1} has at most $4s + 8$ singular values not equal to one and hence GC^{-1} has singular values clustered around one.*

Proof. We first note that

$$GC^{-1} = I + (G - C)C^{-1} \equiv I + L_1,$$

where $\text{rank}(L_1) \leq 2s + 4$ by Lemma 6.4.1. Therefore

$$C^{-*}G^*GC^{-1} - I = L_1^*(I + L_1) + L_1,$$

is a matrix of rank at most $2(2s + 4) = 4s + 8$. $\square$

In order to show the fast convergence of preconditioned conjugate gradient type methods with our preconditioner C, one still needs to estimate $\sigma_{\min}(GC^{-1})$, the smallest singular value of GC^{-1}. Similar to the case in Chapter 3, one may consider the following regularized equation:

$$C^{-*}(G^*G + n^{-4-\beta}I)C^{-1}\mathbf{x} = C^{-*}G^*C^{-1}\mathbf{b}. \tag{6.19}$$

The above regularized equation still has singular values clustered around one. Moreover, the error due to regularization is small and tends to zero as n tends to infinity. By using the techniques discussed in Chapters 3 and 5, we have the following lemma and propositions.

Lemma 6.4.2. *Suppose the parameters $\lambda, \mu, \sigma, \gamma, s$, and r are fixed independent of n. Then there exist positive constants τ_1 and τ_2 independent of n such that $\tau_1 \leq \|C^{-1}\|_2 \leq \tau_2 n^2$.*

From Lemma 6.4.2, we remark that our preconditioner C will never be ill-conditioned even if G is ill-conditioned.

Proposition 6.4.2. *Let the parameters* $\lambda, \mu, \sigma, \gamma, s,$ *and* r *are fixed indepen-dent of* n, *then for any positive* β, *the regularized preconditioned matrix*

$$C^{-*}(G^*G + n^{-4-\beta}I)C^{-1} \tag{6.20}$$

has eigenvalues clustered around one and the smallest eigenvalue decreases at a rate no faster than $O(n^{-4-\beta})$. *Furthermore, the error introduced by the regularization is of the order* $O(n^{-\beta})$.

Thus we conclude that PCG type methods applied to preconditioned sys-tem (6.19) with $\beta > 0$ will converge in at most $O(\log n)$ steps; see Chan [19, Lemma 3.8.1]. To minimize the error introduced by the regularization, one can choose a large β. Recall that regularization is required only when the smallest singular value of the matrix GC^{-1} tends to zero faster than $O(n^{-\alpha})$ for any $\alpha > 0$. In general we do not know if G is ill-conditioned, i.e. $\sigma_{min}(G^*G)$ decreases like $O(e^{-n})$. However, we may always assume this is the case and apply the CG method to the regularized equation (6.19). From Proposition 6.4.2, the error in 2-norm due to the regularization (cf. Lemma 6.4.2) is given by

$$||C^{-*}C^{-1}||_2 n^{-4-\beta} \leq ||C^{-*}||_2 ||C^{-1}||_2 n^{-4-\beta} \leq \tau_2^2 n^{-\beta}. \tag{6.21}$$

The error can be reduced as small as possible by choosing a large β. We remark, however, that in all our numerical tests in Section 6.6, we found that there is no need to add the regularization.

6.5 Computational and Memory Cost Analysis

In this section, we give the computational cost of our PCG method. From (6.14) and (6.17), the construction of our preconditioner C has no cost. The main computational cost of our method comes from the matrix vector mul-tiplication of the form $G\mathbf{x}$, and solving the preconditioner system $C\mathbf{y} = \mathbf{r}$. By making use of the band structure of G, the matrix vector multiplication $G\mathbf{x}$ can be done in $O((s+1)n)$ operations. The solution for $C\mathbf{y} = \mathbf{r}$ can be written as follows (cf. (6.18)):

$$\mathbf{y} = (I_{s+1} \otimes F_n) \cdot P \cdot \text{Diag}(\tilde{C}_1^{-1}, C_2^{-1}, \ldots, C_n^{-1}) \cdot P^t \cdot (I_{s+1} \otimes F_n^*)\mathbf{r}. \tag{6.22}$$

The matrix vector multiplication of the forms $F_n\mathbf{x}$ and $F_n^*\mathbf{x}$ can be done in $O(n \log n)$ operations. By using the following lemma, the solution of the linear system

$$\text{Diag}(\tilde{C}_1^{-1}, C_2^{-1}, \ldots, C_n^{-1})\mathbf{y} = \mathbf{b} \tag{6.23}$$

can be obtained in $O((n+1)n)$ operations. Hence the cost of solving (6.22) is $O((s+1)n \log n + (s+1)n)$.

We conclude that in each iteration of the PCG method, we need $O((s + 1)n \log n)$ operations. The cost per iteration of the Block Gauss-Seidel (BGS) method is of $O((s + 1)n)$. This can be done by making use of the band structure of the diagonal blocks of the generator matrix A. Although the PCG method requires an extra $O(\log n)$ operations in each iteration, the fast convergence rate of our method can more than compensate for this minor overhead (see the numerical results in Section 6.6). In Propositions 6.4.1 and 6.4.2, we proved that the preconditioned linear system and the preconditioned linear system with regularization (6.20) has singular values clustered around one. Numerical results indicated that the number of iterations for convergence is roughly constant, so that the total number of operations for solving the steady-state probability vector is roughly $O((s + 1)n \log^2 n)$. Both the PCG method and the BGS method require $O((s + 1)n)$ memory. Clearly at least $O((s + 1)n)$ memory is required to store the approximated solution in each iteration.

6.6 Numerical Examples

Here we employ the Conjugate Gradient Squared (CGS) method to solve the preconditioned system. We compare our PCG method with the classical iterative method, the Block Gauss-Seidel (BGS) in the following numerical examples. In our examples, we let

$$\lambda = \mu = \gamma = \sigma = 1.$$

The stopping criterion for both PCGS and BGS is

$$\|A\mathbf{p}_k\|_2 < 10^{-12},$$

where $\mathbf{p}_k$ is the approximated solution obtained at the kth iteration. The initial guess for both methods is the unit vector $\mathbf{e}_1 = (1, 0, \ldots, 0)^t$. We give the number of iterations for convergence for fixed number of machines (Table 6.1) and fixed number of maintenance facility (Table 6.2). for different values of $n = m + h + 1$. The symbols I, C, BGS represent the methods used, namely, CGS without preconditioner, CGS with preconditioner C, and the BGS method. We see that the number of iterations for convergence is roughly constant independent of n. The symbol "**" signifies the number of iterations is greater than 1000. We conclude that among all the tested methods, our proposed method is the best.

6.7 System Performance Analysis

In this section, we are going to illustrate some applications of the proposed model for the FMS. In the following numerical examples, for simplicity, we

Table 6.1. number of iterations for convergence ($s = 2$)

	$r = 1$			$r = 2$		
n	I	C	BGS	I	C	BGS
16	41	6	323	83	6	503
64	185	7	**	109	7	**
256	**	7	**	**	9	**
1024	**	7	**	**	9	**

Table 6.2. Number of iterations for convergence ($r = 1$)

	$s = 4$			$s = 8$		
n	I	C	BGS	I	C	BGS
16	74	8	379	118	14	332
64	**	10	**	**	20	**
256	**	11	**	**	22	**
1024	**	11	**	**	22	**

consider an FMS in which backlog is not permitted. We let the demand arrival rate λ be 3, the production rate μ of each machine be 1 and the machine repairing rate σ be 1. We also fix the unit inventory cost C_I, operating cost C_O per unit time, repairing cost C_R per unit time, and the profit per unit product C_P to be $5, 5, 20$, and 60 respectively.

In the first example, we demonstrate the reliability of machine is an important factor in the performance of an FMS.

Example 6.7.1. We consider the system performance by varying the machine breaking down rate γ. We test three different values of γ. The first case is $\gamma = 0.01$ which represents the case that the machine seldom breaks down and is highly reliable. The second case is $\gamma = 0.1$; it represents that machine is moderately reliable. The third case is $\gamma = 1$; the machine is highly unreliable and breaks down very often. We fix the number of machines $s = 4$ and vary the number of maintenance facility r from 1 to 4. Using our proposed model and the numerical algorithm, we compute the following tuples (h, TH, IT, PT) (recall (6.5) and (6.9)) for the mentioned values of r and γ. Here h is the optimal hedging point, TH and IT are respectively the corresponding system throughput and the percentage of machine idle time, and PT is the optimal average profit under optimal HPP policy. From the numerical results and many other tested numerical examples, we observe that under optimal HPP, for a given maintenance facility r, there are large deviations in average profit and machine idle time for different machine reliability γ (Table 6.3). Thus the reliability of machines should be taken in account in the FMS modeling. We also observe that when the machines are highly

Table 6.3. The performance of FMS for different machine reliability

r	$\gamma = 0.01$	$\gamma = 0.1$	$\gamma = 1$
1	(5, 2.78, 30.1%, 130.0)	(6, 2.80, 25.7%, 125.1)	(15, 1.80, 0.2%, 45.2)
2	(5, 2.78, 30.1%, 130.0)	(6, 2.81, 26.1%, 126.1)	(10, 2.41, 4.1%, 85.3)
3	(5, 2.78, 30.1%, 130.0)	(6, 2.81, 26.1%, 126.1)	(9, 2.49, 5.9%, 91.3)
4	(5, 2.78, 30.1%, 130.0)	(6, 2.80, 26.1%, 126.1)	(9, 2.50, 6.0%, 91.8)

reliable ($\gamma = 0.01$), the number of maintenance facilities can be kept at a minimum level. However, when the machines are highly unreliable ($\gamma = 1$), the number of maintenance facilities available is an important factor for the system performance. Furthermore, we also observed that the more reliable the machines are, the less inventory we need to keep in the system.

The second example discusses the design of the FMS.

Example 6.7.2. Suppose that in the FMS, there can be only one maintenance facility, i.e. $r = 1$. Moreover, due to limited capital, at most four machines can be implemented in the system and each machine has a failure rate γ of 1. Assuming the other system parameters are kept the same as in Example 6.7.1. What is the optimal number of machines to be placed in the system such that the average profit is maximized? Again we use our proposed model and the numerical algorithm to compute the following results (Table 6.4). In this case, the optimal number of machines to be placed in the system is three.

Table 6.4. Performance of FMS for different number of machines

$s = 1$	(28, 0.7, 0.0%, 28.5)
$s = 2$	(22, 1.2, 0.0%, 46.2)
$s = 3$	(18, 1.6, 0.1%, 51.5)
$s = 4$	(15, 1.8, 0.2%, 42.2)

In the third example we consider the problem of minimum maintenance facility.

Example 6.7.3. Suppose there are eight ($s = 8$) moderately reliable ($\gamma = 0.1$) machines in the FMS. Assuming the other system parameters are kept the same as in the first example, what is the optimal number of maintenance facility to be placed in the system? In this case, the minimum number of maintenance facilities to be placed in the system is three (Table 6.5). A further increase in maintenance facilities does not improve the average profit and the performance of the FMS.

Table 6.5. Performance of FMS for different number of maintenance.

$r = 1$	(3, 2.9, 61.0%, 112.2)
$r = 2$	(3, 2.9, 61.9%, 115.0)
$r = 3$	(3, 2.9, 61.9%, 115.1)
$r = 4$	(3, 2.9, 61.9%, 115.1)
$r = 5$	(3, 2.9, 61.9%, 115.1)
$r = 6$	(3, 2.9, 61.9%, 115.1)
$r = 7$	(3, 2.9, 61.9%, 115.1)
$r = 8$	(3, 2.9, 61.9%, 115.1)

6.8 Summary and Discussion

A Markovian queuing model has been proposed for flexible manufacturing systems of multiple unreliable machines under hedging point production policy. A numerical algorithm based on the preconditioned conjugate gradient method is presented to solve the steady-state probability distribution. A convergence rate proof of our method is also given when the size of inventory levels is large. Numerical examples are given to demonstrate that the machine reliability and maintainability have important effects on the performance of the system. Other applications of the model are also illustrated.

Our proposed model can still cope with the case when machines are not identical. It is interesting to extend our model to non-Markovian repairing and production processes. For example, our model can be extended to handle the case when the repairing process is a sequence of exponential distributed repairing steps.

Exercises

6.1 Consider a manufacturing system of m identical unreliable machines producing one type of product. Each machine can produce at a rate of d when it is normal. The normal time is exponentially distributed with mean time λ^{-1}. When a machine is broken, it is subject to an l-phase repairing process (see Fig. 6.1). In each phase, the repairing time is exponentially distributed with parameter $\mu_i(r)(i = 1, 2, \ldots, l)$ and r is the number of operators in the system. At any time, there is at most one machine in the repairing process. The repairing time of each phase depends on the number of operators.

Let $N(t)$ be the number of broken machines and $X(t)$ be the phase of the current repair if any of the machines are broken, and $X(t) = 0$ otherwise, at time t.

(a) Explain why $\{(N(t), X(t)), t \geq 0\}$ is a continuous time Markovian process on the state space

$$S = \{(0,0), (i,j), \ i = 1, 2, \ldots, m; \ j = 1, 2, \ldots, l\}.$$

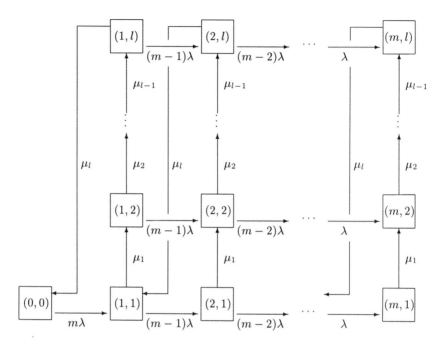

Fig. 6.1. The states (i,j) of the one-queue repairing model.

(b) Write down the generator matrix in matrix form.

(c) Derive an algorithm (direct method) to obtain the steady-state probability distribution in $O(lm)$ operations.

7. Manufacturing Systems of Two Machines in Tandem

7.1 Introduction

In this chapter, we study a two-stage manufacturing system. The system consists of two reliable machines (one in each stage) producing one type of product. Each product has to go through the manufacturing processes in the two stages before it is finished. We assume an infinite supply of raw material for the manufacturing process at the first machine. The mean processing time for one unit of product in the first and second machine are exponentially distributed with parameters μ_1^{-1} and μ_2^{-1} respectively. A buffer B_1 of size b_1 is placed between the two machines to store the products which are finished by the first machine and waiting for further operations by the second machine. Finished products are put in a buffer B_2 of maximum size b_2. The inter-arrival time of a demand is assumed to be exponentially distributed with parameter λ^{-1}. A finite backlog of finished product is allowed in the system. The maximum allowable backlog of product is m. When the inventory level of the finished product is $-m$, any arrival demand will be rejected. Hedging Point Production (HPP) policy is employed as the inventory control in both buffers B_1 and B_2. We focus on HPP policies for our manufacturing systems.

For the first machine, the hedging point is b_1 and the inventory level is non-negative. For the second machine, the hedging point is h; however, the inventory level of buffer B_2 can be negative, because we allow a maximum backlog of m. Very often the buffer size and the inventory levels of B_2 are much larger than that of B_1. This is because the finished product is usually more preferable than the partially finished product in a manufacturing system. Inventory cost and backlog cost can be written in terms of the steady-state probability of the system.

The machine-inventory system is modeled as a Markov chain problem. It turns out that the process is an irreducible continuous time Markov chain. We give the generator matrix for the process, and the PCG method is employed to compute the steady-state probability distribution. Preconditioners are constructed by taking a *circulant approximation* of the generator matrix of the system. We will show that if the parameters μ_1, μ_2, λ, and b_1 are fixed and independent of $n = m + h + 1$, then the preconditioned linear system has singular values clustered around one as n tends to infinity. Hence the CG methods will converge very fast when applied to solving the preconditioned linear system. Numerical examples are given in Section 7.6 to verify our claim.

The remainder of this chapter is organized as follows. In Section 7.2, we formulate the manufacturing system and give the generator matrix of the corresponding continuous time Markov chain. In Section 7.3, a preconditioner is constructed for the generator matrix. In Section 7.4, we prove that the preconditioned linear system has singular values clustered around one. In Section 7.5, we give a cost analysis for our method. In Section 7.6, numerical examples are given to demonstrate the fast convergence rate of our method. An approximation method for solving the steady-state probability vector is also discussed in Section 7.7. A summary is given in Section 7.8 to address further extensions of our model.

7.2 The Two-Machine Manufacturing System

In this section, we construct the generator matrix for the manufacturing system. Let us define the following system parameters:

(i) λ^{-1}, the mean inter-arrival time of a demand,
(ii) μ_1^{-1}, the mean unit processing time of the first machine,
(iii) μ_2^{-1}, the mean unit processing time of the second machine,
(iv) b_1, buffer size for the first machine,
(v) b_2, maximum buffer size for the finished products,
(vi) h, the hedging point,
(vii) m, the maximum allowable backlog,
(viii) c_{I_1}, unit inventory cost for the first buffer B_1,
(ix) c_{I_2}, unit inventory cost for the second buffer B_2,
(x) c_B, unit backlog cost of the finished products.

We note that the inventory level of the first buffer cannot be negative or exceed the buffer size b_1. Thus the total number of inventory levels in the first buffer is $b_1 + 1$. For the second buffer, under the HPP policy, the maximum possible inventory level is $h(h \leq b_2)$; see Fig. 7.1. Since we allow a maximum backlog of m in the system, the total number of possible inventory levels in the second buffer is $n = m + h + 1$. In practice the value of n can easily go up to thousands. We let $z_1(t)$ and $z_2(t)$ be the inventory levels of the first and second buffer at time t respectively. Then $z_1(t)$ and $z_2(t)$ take integer values in $[0, b_1]$ and $[-m, h]$ respectively. Thus the joint inventory process

$$\{(z_1(t), z_2(t)), t \geq 0\}$$

is a continuous time Markov chain taking values in the state space

$$S = \{(z_1(t), z_2(t)) : z_1 = 0, \ldots, b_1, \ z_2 = -m, \ldots, h.\}.$$

The steady-state probability distribution of the system is given by

Fig. 7.1. Manufacturing systems of two machines in tandem.

$$\lim_{t\to\infty} \text{Prob}\{(z_1(t), z_2(t)) = (i,j)\} = p(i,j), \quad i = 0,\ldots,b_1; \; j = -m,\ldots,h.$$

We order inventory states lexicographically, according to z_1 first and then z_2, and the tri-diagonal block generator for the joint inventory system can be obtained as follows:

$$A = \begin{pmatrix} \Lambda + \mu_1 I_n & \Sigma & & & 0 \\ -\mu_1 I_n & \Lambda + D + \mu_1 I_n & \Sigma & & \\ & \ddots & \ddots & \ddots & \\ & & -\mu_1 I_n & \Lambda + D + \mu_1 I_n & \Sigma \\ 0 & & & -\mu_1 I_n & \Lambda + D \end{pmatrix}, \quad (7.1)$$

where

$$\Lambda = \begin{pmatrix} 0 & -\lambda & & 0 \\ \lambda & \ddots & & \\ & & \ddots & -\lambda \\ 0 & & & \lambda \end{pmatrix}, \quad (7.2)$$

$$\Sigma = \begin{pmatrix} 0 & & & 0 \\ -\mu_2 & \ddots & & \\ & \ddots & \ddots & \\ 0 & & -\mu_2 & 0 \end{pmatrix}, \quad (7.3)$$

I_n is the $n \times n$ identity matrix and D is the $n \times n$ diagonal matrix

$$D = \text{Diag}(\mu_2,\ldots,\mu_2,0). \quad (7.4)$$

We are interested in solving the steady-state probability distribution **p** of the generator matrix A. Useful quantities such as the throughput of the system

$$\left(1 - \sum_{j=-m}^{h} p(0,j)\right)\mu_2 \quad (7.5)$$

and mean number of products in buffers B_1 and B_2 (work-in-process)

$$\sum_{i=1}^{b_1} \left(\sum_{j=-m}^{h} p(i,j) \right) i \quad \text{and} \quad \sum_{j=1}^{h} \left(\sum_{i=0}^{b_1} p(i,j) \right) j \tag{7.6}$$

can be written in terms of $\mathbf{p}$; see for instance [38, 41, 45, 106]. Furthermore the average running cost and the average profit of the system can be also written in terms of $\mathbf{p}$,

$$C_{I_1} \sum_{i=1}^{b_1} \left(\sum_{j=-m}^{h} p(i,j) \right) i + C_{I_2} \sum_{j=1}^{h} \left(\sum_{i=0}^{b_1} p(i,j) \right) j + C_B \sum_{j=1}^{m} \left(\sum_{i=0}^{b_1} p(i,-j) \right) j.$$

We note that the generator A is irreducible, has zero column sum, positive diagonal entries and non-positive off-diagonal entries, so that A has a one-dimensional null space with a right positive null vector. The steady-state probability distribution $\mathbf{p}$ is then equal to the normalized form of the positive null vector. Similar to the previous chapters, we consider the following equivalent linear system

$$G\mathbf{x} \equiv (A + \mathbf{e}_1 \mathbf{e}_1^t)\mathbf{x} = \mathbf{e}_1, \tag{7.7}$$

where $\mathbf{e}_1 = (1,0,\ldots,0)^t$ is the $(b_1+1)(m+h+1)$ unit vector. Similarly one can show that the following linear system (7.7) is non-singular and hence the steady-state probability distribution can be obtained by normalizing the solution of (7.7). Clearly we have the following lemma.

Lemma 7.2.1. *The matrix G is non-singular.*

However, the analytical solution of $\mathbf{p}$ is not generally available. Therefore, most of the techniques employed for the analysis are analytical approximations and numerical solutions. Yamazaki [114] gave an analytical approximation for the two-station queuing model under the assumption of infinite buffer size. Our manufacturing system deals with the realistic situation of finite buffer size. Usually, by making use of the block structure of the generator matrix A, a classical iterative method such as block Gauss-Seidel is applied in solving the steady-state probability distribution to save computational cost; see Siha [106]. However, in general the classical iterative methods have slow convergence rates; see the numerical results in Section 7.6. We employ Conjugate Gradient (CG) type methods to solve the preconditioned system.

7.3 Construction of Preconditioners

In this section, we construct a preconditioner by taking the circulant approximation of blocks Λ, Σ and D of A. It is well known that any $n \times n$ circulant matrix C_n is characterized by its first column (or the first row) and can be diagonalized by the discrete Fourier matrix F_n, i.e. $C_n = F_n \Omega_n F_n^*$, where F_n^* is the conjugate transpose of F_n and Ω_n is a diagonal matrix containing the eigenvalues of C_n. The matrix vector multiplication of the forms $F_n \mathbf{y}$ and $F_n^* \mathbf{y}$ can be obtained in $O(n \log n)$ operations by the Fast Fourier Transform (FFT). By completing Λ, Σ and D to circulant matrices, we define the circulant approximation $c(\Lambda)$, $c(\Sigma)$ and $c(D)$ as follows:

$$c(\Lambda) = \begin{pmatrix} \lambda & -\lambda & & 0 \\ & \lambda & \ddots & \\ & & \ddots & -\lambda \\ -\lambda & & & \lambda \end{pmatrix}, \tag{7.8}$$

$$c(\Sigma) = \begin{pmatrix} 0 & & & -\mu_2 \\ -\mu_2 & \ddots & & \\ & \ddots & \ddots & \\ 0 & & -\mu_2 & 0 \end{pmatrix} \tag{7.9}$$

and

$$c(D) = \mathrm{Diag}(\mu_2, \ldots, \mu_2, \mu_2).$$

From (7.8) we define the circulant approximation $c(A)$ of the generator matrix A as follows: $c(A) = \mu_1 I_n +$

$$\begin{pmatrix} c(\Lambda) & c(\Sigma) & & & 0 \\ -\mu_1 I_n & c(\Lambda) + c(D) & c(\Sigma) & & \\ & \ddots & \ddots & \ddots & \\ & & -\mu_1 I_n & c(\Lambda) + c(D) & c(\Sigma) \\ 0 & & & -\mu_1 I_n & c(\Lambda) + c(D) - \mu_1 I_n \end{pmatrix}. \tag{7.10}$$

From (7.8) and Davis [62], we have the following lemmas.

Lemma 7.3.1. $\mathrm{Rank}(c(\Lambda) - \Lambda) = \mathrm{Rank}(c(\Sigma) - \Sigma) = \mathrm{Rank}(c(D) - D) = 1$.

Lemma 7.3.2. *The matrices $c(\Lambda)$ and $c(\Sigma)$ can be diagonalized by the discrete Fourier transform F_n. The eigenvalues of $c(\Lambda)$ and $c(\Sigma)$ are given by*

$$F_n^* c(\Lambda) F_n = \mathrm{Diag}(\nu_1, \nu_2, \ldots, \nu_n)^t$$

and

$$F_n^* c(\Sigma) F_n = \mathrm{Diag}(\xi_1, \xi_2, \ldots, \xi_n)^t,$$

where

$$\begin{cases} \nu_j = \lambda(1 - e^{\frac{2\pi(j-1)}{n} i}), & j = 1, 2, \ldots, n, \\ \xi_j = -\mu_2 e^{\frac{2\pi(j-1)}{n} i}, & j = 1, 2, \ldots, n. \end{cases} \tag{7.11}$$

There exists a permutation matrix P such that

$$P^t \cdot (I_{b_1+1} \otimes F_n^*) \cdot c(A) \cdot (I_{b_1+1} \otimes F_n) \cdot P = \text{Diag}(C_1, C_2, \ldots, C_n),$$

where

$$C_i = \begin{pmatrix} \mu_1 + \nu_i & \xi_i & & & & 0 \\ -\mu_1 & \mu_1 + \mu_2 + \nu_i & \xi_i & & & \\ & \ddots & \ddots & \ddots & & \\ & & -\mu_1 & \mu_1 + \mu_2 + \nu_i & \xi_i \\ 0 & & & -\mu_1 & \mu_2 + \nu_i \end{pmatrix}. \qquad (7.12)$$

We note that all C_i except C_1 are strictly diagonal dominant and therefore they are non-singular. By similar argument in the proof of Lemma 7.2.1,

$$\tilde{C}_1 = (C_1 + \mathbf{f}\mathbf{f}^t) \qquad (7.13)$$

is non-singular where $\mathbf{f} = (1, 0, \ldots, 0)$. We define the preconditioner C as

$$C = (I_{b_1+1} \otimes F_n) \cdot P \cdot \text{Diag}(\tilde{C}_1, C_2, \ldots, C_n) \cdot P^t \cdot (I_{b_1+1} \otimes F_n^*). \qquad (7.14)$$

7.4 Convergence Analysis

In this section, we study the convergence rate of the PCG method when $n = m + h + 1$ is large. In practice, the number of possible inventory states n is much larger than b_1 in the manufacturing systems and can easily go up to thousands. We prove that if all parameters μ_1, μ_2, λ, and b_1 are fixed independent of n, then the preconditioned system GC^{-1} has singular values clustered around one as n tends to infinity. Hence when CG type methods are applied to solving the preconditioned system, we expect fast convergence. Numerical examples are given in Section 7.6 to verify our claim. We begin the proof with the following lemma.

Lemma 7.4.1. We have $\text{rank}(G - C) \leq 2(b_1 + 2)$.

Proof. From (7.7), we have

$$\text{rank}(G - A) = 1,$$

and from Lemma 7.3.1, we see that

$$\text{rank}(A - c(A)) = 2(b_1 + 1).$$

Using (7.12) and (7.14), we have $c(A)$ and C differ by a rank one matrix. Therefore, we have

$$\begin{aligned} \text{rank}(G - C) &\leq \text{rank}(G - A) + \text{rank}(A - c(A)) + \text{rank}(c(A) - C) \\ &= 2(b_1 + 2). \end{aligned}$$

$\square$

Proposition 7.4.1. *The preconditioned matrix GC^{-1} has at most $4(b_1 + 2)$ singular values not equal to one. Hence GC^{-1} has singular values clustered around one when n tends to infinity.*

Proof. We first note that

$$GC^{-1} = I + (G - C)C^{-1} \equiv I + L_1,$$

where $\text{rank}(L_1) \leq 2(b_1 + 2)$ by Lemma 7.4.1. Therefore

$$C^{-*}G^*GC^{-1} - I = L_1^*(I + L_1) + L_1,$$

is a matrix of rank at most $4(b_1 + 2)$. Thus the number of singular values of GC^{-1} that are different from one is a constant independent of n. Hence the preconditioned matrix $C^{-1}G$ has singular values clustered around one. $\square$

We remark that if the preconditioned matrix is ill-conditioned, the regularization technique discussed in Chapter 3 can be applied to solve the problem. But in the tested numerical examples, regularization is not necessary.

7.5 Computational Cost Analysis

From (7.10) and (7.12), the construction of our preconditioner C need no cost. The main computational cost of our method comes from the matrix vector multiplication of the form Gx, and solving the preconditioner system $Cy = r$. By making use of the band structure of G, the matrix vector multiplication Gx can be done in $O((b_1 + 1)n)$ operations. The solution for $Cy = r$ can be written as follows (cf. 7.14):

$$\mathbf{y} = (I_{b_1+1} \otimes F_n) \cdot P \cdot \text{Diag}(\tilde{C}_1^{-1}, C_2^{-1}, \ldots, C_n^{-1}) \cdot P^t \cdot (I_{b_1+1} \otimes F_n^*)\mathbf{r}. \quad (7.15)$$

The matrix-vector multiplication of the forms $F_n x$ and $F_n^* x$ can be done in $O(n \log n)$ operations. The solution of the linear system

$$\text{Diag}(\tilde{C}_1^{-1}, C_2^{-1}, \ldots, C_n^{-1})\mathbf{y} = \mathbf{b} \quad (7.16)$$

can be obtained in $O((b_1 + 1)n)$ operations. Hence the cost for solving (7.15) is $O((b_1 + 1)n \log n + (b_1 + 1)n)$. We conclude that in each iteration of the PCG method, we need $O((b_1 + 1)n \log n)$ operations. The cost per iteration of the Block Gauss-Seidel (BGS) method is $O((b_1 + 1)n)$. This can be done by making use of the band structure of the diagonal blocks of the generator matrix A. Although the PCG method requires an extra $O(\log n)$ operations in each iteration, the fast convergence rate (roughly constant independent of n) of our method can more than compensate for this minor overhead (see the numerical examples in Section 7.6). In Proposition 7.4.1, we proved that the preconditioned linear system has singular values clustered around one, so we expect the PCG method to converge very fast. The convergence rate of BGS is roughly linear in n. Thus the total cost for the PCG method and the BGS method are

$$O((b_1 + 1)n \log n) \quad \text{and} \quad O((b_1 + 1)n^2)$$

respectively. Both PCG and BGS require $O((b_1 + 1)n)$ memory. Clearly at least $O((b_1 + 1)n)$ memory is required to store the approximated solution in each iteration.

7.6 Numerical Examples

We apply the Conjugate Gradient Squared (CGS) method to solve the preconditioned system. We compare our PCG method with the BGS method in the following numerical examples. In the examples, we let

$$\lambda = 1, \mu_1 = 3/2, \text{ and } \mu_2 = 3.$$

The stopping criterion for both PCGS and BGS is

$$\|A\mathbf{p}_k\|_2 < 10^{-12},$$

where $\mathbf{p}_k$ is the approximated solution obtained at the kth iteration. The initial guess for both methods is the unit vector $\mathbf{e} = (1, 0, \ldots, 0)^t$. All the computations are done in a HP 712/80 workstation with MATLAB. We give the number of iterations for convergence of PCGS and BGS (Table 7.1) for different values of b_1. The symbols I, C, BGS represent the methods used, namely, CGS without preconditioner, CGS with preconditioner C, and the BGS method. The symbol "**" signifies that the number of iterations is greater than 200.

Table 7.1. Number of iterations for PCG and BGS

n	$b_1 = 1$			$b_1 = 2$			$b_1 = 4$			$b_1 = 8$		
	I	C	BGS	I	C	BGS	I	C	BGS	I	C	BGS
16	34	5	72	46	8	71	54	11	71	64	19	72
64	129	7	142	130	8	142	130	11	142	139	19	142
256	**	8	**	**	8	**	196	11	**	**	19	**
1024	**	8	**	**	8	**	**	11	**	**	19	**

7.7 Approximation Method

Even through our method can be generalized to handle machines in tandem (see Ching [38, 41]) i.e. a l-stage production process, one should note that the number of states in the system grows exponentially with respect to the number of stages l. Therefore seeking for an approximation of the steady-state probability distribution is natural. In the following, we are going to obtain an approximate solution of the system steady-state probability distribution. For the first machine M_1, one may approximate the inventory levels in B_1 by the simple Markovian one queue network with "service rate" and "arrival rate" being μ_1 and μ_2 respectively. Thus the steady-state distribution of the inventory levels in buffer B_1 will be given by the following standard result (see Chapter 1):

$$p_k = \frac{\rho^{b_1-k}(1-\rho)}{1-\rho^{b_1+1}}, \quad k = 0, 1, \ldots, b_1, \tag{7.17}$$

where p_k is steady-state probability that the inventory level in B_1 is k and $\rho = \mu_2/\mu_1$.

For the second machine M_2, we note that it can only produce product at an average rate μ_2 when buffer B_1 is non-empty. Therefore there are two important states of buffer B_1, empty and non-empty. Their steady-state probabilities are of course given by

$$p_0 = \frac{\rho^{b_1}(1-\rho)}{1-\rho^{b_1+1}} \quad \text{and} \quad 1 - p_0 = \frac{1-\rho^{b_1}}{1-\rho^{b_1+1}} \tag{7.18}$$

respectively. We may model these two states by a two-state Markovian process (the MMPP) with the generator matrix H:

$$\begin{array}{l} \text{State 1 : not empty} \\ \text{State 2 : empty} \end{array} \begin{pmatrix} f & -e \\ -f & e \end{pmatrix}.$$

The mean time in the state "empty" should be $1/\mu_1$ and the ratio of the mean duration of two states should equal the ratio of their steady-state probabilities. Therefore we have the following two equations:

$$\frac{1}{e} = \frac{1}{\mu_1} \quad \text{and} \quad \frac{1/e}{1/f} = \frac{p_0}{1-p_0}. \tag{7.19}$$

Solving the two equations we get

$$e = \mu_1 \quad \text{and} \quad f = \frac{\rho^{b_1}(1-\rho)\mu_1}{1-\rho^{b_1}}. \tag{7.20}$$

Thus the generator matrix for the joint process of the states of buffer B_1 (empty and non-empty) and the inventory levels of the buffer B_2 can be written down and is given by the following $2(b_1 + 1) \times 2(b_1 + 1)$ matrix:

$$B = \begin{pmatrix} H+F & -L & & & 0 \\ -F & H+L+F & -L & & \\ & \ddots & \ddots & \ddots & \\ & & -F & H+L+F & -L \\ 0 & & & -F & H+L \end{pmatrix} \qquad (7.21)$$

where

$$F = \begin{pmatrix} \mu_2 & 0 \\ 0 & 0 \end{pmatrix} \quad \text{and} \quad L = \begin{pmatrix} \lambda & 0 \\ 0 & \lambda \end{pmatrix}. \qquad (7.22)$$

The generator matrix B is irreducible and therefore its steady-state probability distribution exists. The steady-state probability distribution can be solved in $O(b_2)$ operations. The performance of the approximation method can be found in Ching [49].

7.8 Summary and Discussion

A two-stage manufacturing system is studied in this chapter. The PCG method is applied to solving the steady-state probability distribution of the system. Preconditioner is constructed by taking the circulant approximation of the generator matrix of the system. Numerical examples are given to demonstrate the fast convergence rate of the method.

One may further consider the case when the manufacturing system has multiple identical machines in each stage (r_1 machines in stage one and r_2 machines in stage two). In this case, the generator matrix $\hat{A}$ will be given as follows (see Siha [106]):

$$\hat{A} = \begin{pmatrix} \Lambda+\Gamma & \Sigma_1 & & & 0 \\ -\Gamma & \Lambda+D_1+\Gamma & \Sigma_2 & & \\ & \ddots & \ddots & \ddots & \\ & & -\Gamma & \Lambda+D_{b_1-1}+\Gamma & \Sigma_{b1} \\ 0 & & & -\Gamma & \Lambda+D_{b_1} \end{pmatrix}, \qquad (7.23)$$

where

$$\Lambda = \begin{pmatrix} 0 & -\lambda & & 0 \\ \lambda & \ddots & \ddots & \\ & \ddots & \ddots & -\lambda \\ 0 & & & \lambda \end{pmatrix}, \qquad (7.24)$$

$$\Sigma_i = \begin{pmatrix} 0 & & & 0 \\ -\min(i,r_2)\mu_2 & \ddots & & \\ & \ddots & \ddots & \\ 0 & & -\min(i,r_2)\mu_2 & 0 \end{pmatrix}, \qquad (7.25)$$

D and Γ are the $n \times n$ diagonal matrices

$$D_i = \min(i, r_2)\mathrm{Diag}(\mu_2, \mu_2, \ldots, \mu_2, 0),$$

and

$$\Gamma = \mathrm{Diag}(\mu_1, 2\mu_1, \cdots, r_1\mu_2, \ldots, r_1\mu_2)$$

respectively. The circulant approximation techniques in Section 7.3 can be applied to the construction of preconditioner $\hat{C}$ for the generator $\hat{A}$. The circulant approximation of Λ, Σ_i, D_i, and Γ_i are then given as follows:

$$c(\Lambda) = \begin{pmatrix} \lambda & -\lambda & & 0 \\ & \lambda & \ddots & \\ & & \ddots & -\lambda \\ -\lambda & & & \lambda \end{pmatrix}, \tag{7.26}$$

$$c(\Sigma_i) = \begin{pmatrix} 0 & & & -r_2\mu_2 \\ -r_2\mu_2 & \ddots & & \\ & \ddots & \ddots & \\ 0 & & -r_2\mu_2 & 0 \end{pmatrix}, \tag{7.27}$$

$$c(D) = \mathrm{Diag}(r_2\mu_2, \ldots, r_2\mu_2),$$

and

$$c(\Gamma) = \mathrm{Diag}(r_1\mu_1, \ldots, r_1\mu_2).$$

Following similar proof in Section 7.4, we can also prove that the preconditioned system $\hat{C}^{-1}\hat{A}$ has singular values clustered around one. Thus the PCG method will converge very fast when apply to solving the preconditioned matrix $\hat{C}^{-1}\hat{A}$ system.

Exercises

7.1 Show that (cf. (7.23)) $\mathrm{rank}(\hat{A} - c(\hat{A}))$ is independent of n.

7.2 Show that the steady-state probability distribution of (7.21) can be solved in $O(b_2)$ operations.

8. Manufacturing Systems with Delivery Time Guarantee

8.1 Introduction

In this chapter, we study production planning in a manufacturing system under *Delivery Time Guarantees* (DTG). DTG policy is a marketing strategy used by many commercial companies to attract and retain customers. Here are some examples of DTG policies used by different companies. United Parcel Services (UPS) guarantees next-day delivery by 8:30 am. Pizza Hut in Hong Kong offers a free pizza if the ordered pizza cannot be served within 20 minutes. Lucky, a major supermarket chain in California, use a "three's a crowd" campaign, which guarantees a new checkout counter will be open if there are more than three people waiting in its checkout queue. Wells Fargo Bank offers a "five minute maximum wait policy" which offers five dollars to the customer if the customer waits for more than five minutes in line. We note that a DTG works under the condition that each customer is given a guaranteed quality of service. But in the mentioned examples if a company fails to fulfill its promise, it will lead to the loss of money, customers, and even the reputation of the company; see [16, 36, 44, 63, 64, 88].

In our discussion, we assume that the guaranteed delivery time T and the unit price P of the product (or service) are the main factors determining the demand rate

$$\lambda = \lambda(P, T)$$

of the product. One standard formulation for the relation (Cobb-Douglas) is given as follows:

$$\lambda = \lambda(P, T) = \sigma P^{-\tau_1} T^{-\tau_2}, \tag{8.1}$$

where τ_1, τ_2 are the price and delivery time guarantee elasticity respectively, and σ is a positive constant; see [69, 74, 82] for instance. In the delivery time guarantee policy, a guaranteed delivery time of T is offered to the customers on each unit of ordered product. Here in our model, we assume that the orders are fulfilled in the first-come-first-served principle. The delivery time consists of two parts: the cycle time and the delivery time. The cycle time is the time between the arrival of an order and the time the requested product leaves the manufacturing system. The delivery time is the time for the requested product to get from the manufacturing system to the customers. Here we model the delivery time by a *Shifted exponential distribution*. Unbiased and consistent estimators are derived for the distribution.

We consider a one-machine manufacturing system. The machine produces one type of product; the inter-arrival time of the demand and the processing time for one unit of product are assumed to be exponentially distributed. Finite backlog of the demands is allowed in the system. A cost of c_B is assigned to each unit of backlog and a cost of c_I is assigned to each unit of inventory. Using the inventory level as an indicator, our interest is to determine the production strategy such that the average profit is maximized. We remark that one may consider optimal production strategy such that the average running cost is minimized.

We employ the (s, S) policy as the production control. An (s, S) policy is characterized by two non-negative integers s and S. The machine keeps on producing until an inventory level of S is reached. Once this inventory level is reached, the production is stopped by shutting down the machine. We let the inventory level to fall to $s(s < S)$, and restart the production. The HPP policy discussed in previous chapters is a particular case of the (s, S) policy with $S = s + 1$. In this chapter, we obtain the analytical form of the steady-state probability distribution of the inventory levels. The average profit of the system can then be written in terms of this probability distribution, and the optimal values of s and S are obtained for any given unit price P of the product and the guaranteed delivery time T.

The remainder of the chapter is organized as follows. In Section 8.2, we present the model of manufacturing system under (s, S) policy. In Section 8.3, we derived unbiased and consistent estimators for the distribution of the delivery time. In Section 8.4, we derive a desirable maximum backlog when the guaranteed delivery time T is given. In Section 8.5, we give the generator matrix for the steady-state probability distribution of the machine-inventory system. The analytical form of this probability distribution is obtained in Section 8.6. In Section 8.7, the average profit of the systems is written in terms of the steady-state probability distribution. In Section 8.8, we consider the case when infinite backlog is allowed. Under some conditions (specified later), we prove that the optimal (s, S) policy is of hedging point type, i.e. $S = (s + 1)$. Numerical examples are given in Section 8.9. Finally a summary is given in Section 8.10 to discuss some possible extensions of the model.

8.2 The Manufacturing System

In this section, we present a one-machine manufacturing system with product delivery time guarantee. We will use the following notation:

(i) μ^{-1}, the mean processing time for a demand for one unit of the product,
(ii) λ^{-1}, the mean inter-arrival time for one unit of the product,
(iii) O, the operating cost for one unit of the product,

(iv) P $(P > O)$, the unit price of the product,

(v) T, the guarantee time of delivery,

(vi) $s < S$, the parameters of the (s, S) policy,

(vii) c_I, the inventory carrying cost per unit of the product (dollars),

(viii) c_P, the penalty cost per unit of the product (dollars),

(ix) c_B, the backlog cost per unit of the product (dollars),

(x) m, the maximum allowable backlog,

(xi) $I_{\max}$, the maximum allowable inventory.

We employ (s, S) policy as the production control. The processing time for one unit of product by the machine and the inter-arrival time for one unit of demand are assumed to be exponentially distributed. The demand rate of the product depends on the unit price P and the guaranteed delivery time T, and is given by (8.1). Backlog of products is allowed in the system. We reject any arrival demand when the inventory level is $-m$, i.e. our maximum backlog capacity is m. A penalty cost of c_P is assigned to each unit of rejected demand. To determine the optimal values of s and S, we have to set an appropriate guaranteed delivery time. In the next section, we study the delivery time distribution.

8.3 The Delivery Time Distribution

In this section, we study the delivery time distribution of an order. The total lead time of an order equals the sum of the cycle time and the delivery time of the product. The processing time of an product is assumed to be exponentially distributed with parameter μ. If an arrival order is not a backlog, then the cycle time (the time between the arrival of order and the time the requested product leaves the manufacturing system) is zero. But if the order is in the jth backlog then the probability density function of the cycle time of the order is the Erlangian distribution of j-phase:

$$f_j(t) = \frac{\mu(\mu t)^j e^{-\mu t}}{j!}. \tag{8.2}$$

Usually the delivery time is greater than a minimum time T_0. The time T_0 can be the packing time of the products or the waiting time for transportation schedule. In our study, we model the delivery time by a "shifted" exponential distribution of parameters T_0 and β:

$$T(t) = \begin{cases} \beta e^{-\beta t} & \text{if} \quad t \geq T_0 \\ 0 & \text{if} \quad t < T_0. \end{cases} \tag{8.3}$$

A similar formulation has been used by Karmarkar [82] in the fractile estimation. In the following, we derive unbiased and consistent estimators for T_0 and β^{-1} respectively. The estimators here are not unique; other forms of unbiased and consistent estimators are available, see Hogg and Craig [75] for instance.

We first let

$$\{T_1, T_2, \ldots, T_n\}$$

be a sample of n delivery times, $\bar{T}$ be the mean of the sample, and $T_{(1)} = \min\{T_i\}$. We have the following lemma for the ordered statistic T_1.

Lemma 8.3.1. *Let* $T_{(1)} \leq T_{(2)} \leq \ldots \leq T_{(n)}$ *be nth-order statistics generated by the exponential distribution*

$$f(t) = \beta e^{-\beta t},$$

then the probability density function for $T_{(1)}$ *is given by*

$$g(t) = n\beta e^{-n\beta t}.$$

Therefore

$$\mathrm{E}(T_{(1)}) = \frac{1}{n\beta} \quad \text{and} \quad \mathrm{Var}(T_{(1)}) = \frac{1}{(n\beta)^2}.$$

Proof. Since each random variable $T_{(i)}$ is exponentially distributed with parameter β, we have

$$\mathrm{Prob}(X_{(i)} \geq Y) = \int_Y^\infty \beta e^{-\beta t} dt, \quad \text{for } i = 1, 2, \ldots, n. \tag{8.4}$$

Thus if $X_{(1) \geq Y}$ we must have all the $X_{(i)} \geq Y$. Therefore we have

$$\mathrm{Prob}(X_{(1)} \geq Y) = \left(\int_Y^\infty \beta e^{-\beta t} dt \right)^n = (e^{-\beta Y})^n = e^{-n\beta Y}. \tag{8.5}$$

Differentiating both sides with respect to Y, we have the probability density function of $T_{(1)}$ given by

$$g(t) = n\beta e^{-n\beta t}. \tag{8.6}$$

For more details about order statistics distributions, see Hogg and Craig [75]. The probability density function $g(t)$ is still an exponential distribution with parameter $n\beta$. Thus we have

$$E(T_{(1)}) = (n\beta)^{-1} \quad \text{and} \quad Var(T_{(1)}) = (n\beta)^{-2}.$$

$\square$

We may estimate T_0 and β by $T_{(1)}$ and $(\bar{T} - T_{(1)})$ respectively. However, by Lemma 8.3.1, we have

$$E(T_{(1)}) = T_0 + \frac{1}{n\beta}. \tag{8.7}$$

Therefore $T_{(1)}$ is a biased estimator and so is $(\bar{T} - T_{(1)})$. Here according to the bias, we consider the estimators $\hat{T}_0$ and $\hat{\beta}^{-1}$ for T_0 and β^{-1} satisfying the equations:

$$\begin{cases} \hat{T}_0 = T_{(1)} - (n\hat{\beta})^{-1}, \\ \hat{\beta}^{-1} = \bar{T} - \hat{T}_0. \end{cases} \qquad (8.8)$$

Solving the equations, we get

$$\begin{cases} \hat{T}_0 = (nT_{(1)} - \bar{T})/(n - 1), \\ \hat{\beta}^{-1} = n(\bar{T} - T_0)/(n - 1). \end{cases} \qquad (8.9)$$

In the following, we prove that $\hat{T}_0$ and $\hat{\beta}^{-1}$ are unbiased and consistent estimators for T_0 and β^{-1} respectively.

Proposition 8.3.1. *The estimators*

$$\hat{T}_0 = (nT_{(1)} - \bar{T})/(n - 1)$$

and

$$\hat{\beta}^{-1} = n(\bar{T} - T_0)/(n - 1)$$

are unbiased and consistent estimators for T_0 and β^{-1} respectively.

Proof. Since

$$\begin{aligned} E(\hat{T}_0) &= \frac{1}{n - 1}(nE(T_{(1)}) - E(\bar{T})) \\ &= \frac{1}{n - 1}(n(T_0 + \beta/n) - T_0 - \beta) = T_0 \end{aligned}$$

and

$$\begin{aligned} E(\hat{\beta}^{-1}) &= \frac{n}{n - 1}(E(\bar{T}) - E(T_{(1)})) \\ &= \frac{n}{n - 1}(\beta^{-1} + T_0 - T_0 - (n\beta)^{-1}) = \beta^{-1}, \end{aligned}$$

the estimators $\hat{\beta}^{-1}$ and $\hat{T}_0$ are unbiased. Moreover, by Lemma 8.3.1, we have

$$\begin{aligned} \text{Var}(\hat{\beta}^{-1}) &= \frac{n^2}{(n - 1)^2}\left(\text{Var}(\bar{T}) + \text{Var}(T_{(1)})\right) \\ &= \frac{n^2}{(n - 1)^2}\left(\frac{n}{(n\beta)^2} + \frac{1}{(n\beta)^2}\right) \end{aligned}$$

and

$$\begin{aligned} \text{Var}(\hat{T}_0) &= \frac{n^2}{(n - 1)^2}\text{Var}(T_{(1)}) + \frac{1}{(n - 1)^2}\text{Var}(\bar{T}) \\ &= \frac{1}{((n - 1)\beta)^2} + \frac{1}{n(n - 1)\beta}. \end{aligned}$$

Thus

$$\lim_{n \to \infty} \text{Var}(\hat{\beta}^{-1}) = \lim_{n \to \infty} \text{Var}(\hat{T}_0) = 0, \qquad (8.10)$$

and therefore $\hat{\beta}^{-1}$ and $\hat{T}_0$ are consistent estimators.   □

8.4 Determination of Maximum Allowable Backlog

In a manufacturing system of limited production capacity, to guarantee customers a delivery time for the product, there should be an upper limit for the maximum allowable backlog m. Here we propose a simple estimation of m. Suppose we have k backlogged orders, the probability that they can be offered within the guarantee time T should be greater than a predetermined value γ ($0 < \gamma < 1$). The *reliability value* (or the *service level*) γ can be 90% or 95% for example. The following proposition gives an inequality relating the quantities γ, T, and m.

Proposition 8.4.1. *Given the delivery guarantee time T and the reliability value γ, the maximum allowable backlog m is the largest positive integer k which satisfies the inequalities:*

$$
\begin{cases}
(1 - \gamma)e^{\mu(T-T_0)} \geq \displaystyle\sum_{j=0}^{k-1} \frac{(T - T_0)^j(\mu^j - \mu^k(\mu - \beta)^{(j-k)})}{j!} \\
\qquad\qquad + \dfrac{\mu^k e^{-(\beta-\mu)(T-T_0)}}{(\mu - \beta)^k} \quad \text{for } \mu \neq \beta, \\
(1 - \gamma)e^{\mu(T-T_0)} \geq \displaystyle\sum_{j=0}^{k-1} \frac{(\mu(T - T_0))^j}{j!} + \dfrac{\mu^k(T - T_0)^k e^{-\mu(T-T_0)}}{k!} \quad \text{for } \mu = \beta.
\end{cases} \tag{8.11}
$$

Proof. The processing time for one product is exponentially distributed with mean μ^{-1}, therefore the probability density function for the total processing time of k products is given by the Erlangian function

$$
f_k(t) = \frac{\mu(\mu t)^{k-1} e^{-\mu t}}{(k - 1)!}.
$$

The delivery time is modeled by the shifted exponential distribution

$$
g(t) = \begin{cases} \beta e^{-\beta t} & \text{if } t \geq T_0 \\ 0 & \text{if } t < T_0. \end{cases} \tag{8.12}
$$

The parameters β and T_0 are estimated by the unbiased and consistent estimators in Proposition 8.3.1. Hence we have

$$
\int_0^{T-T_0} \frac{\mu(\mu t)^{k-1}}{(k - 1)!} e^{-\mu t}[1 - e^{-\beta(T-T_0-t)}]dt \geq \gamma. \tag{8.13}
$$

By making use of the formula

$$
\int_0^T \frac{\mu(\mu t)^{k-1}}{(k - 1)!} e^{-\mu t} dt = 1 - e^{-\mu T} \sum_{j=0}^{k-1} \frac{(\mu T)^j}{j!}, \tag{8.14}
$$

one can obtain the inequalities (8.11). Therefore the maximum allowable backlog m is the largest integer k satisfying the above inequality. □

8.5 The Inventory Levels

When the desirable guaranteed delivery time T and the unit price P of the product are decided, the maximum allowable backlog m (Proposition 8.4.1) and the demand rate λ (c.f. (8.1)) can be determined. We then employ the (s, S) policy in the production planning of the manufacturing system. We note that under the (s, S) policy, the inventory levels can take integer values in $[-m, S]$. When the inventory level reaches S, the machine is shut off. When the machine is shut off, the inventory levels can take integer values in $[s, S]$ and once it falls to s, the production is restarted. Therefore there are $(m + 2S - s)$ possible states in the inventory system; see Fig. 8.1.

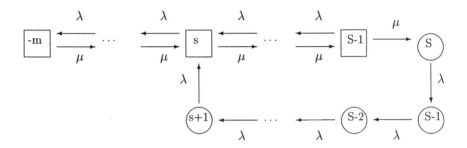

$\bigcirc$ i   The machine is shut off and the inventory level is i.

$\boxed{j}$ The machine is turned on and the inventory level is j.

Fig. 8.1. The machine-inventory system.

We order the states as follows. The first $(S - s)$ states correspond to the situation that the machine is turned off, and the inventory levels are ordered from $(s+1)$ up to S. The next $(m+S)$ states correspond to the situation that the machine is turned on, and we order the inventory levels from $(S-1)$ down to $-m$. We then construct the generator matrix for the machine-inventory system. Under our ordering, we obtain the following generator matrix A for the machine-inventory system:

$$
A = \begin{pmatrix}
\lambda & -\lambda & & & & & & & & & 0 \\
0 & \lambda & -\lambda & & & & & & & & \\
0 & & \ddots & \ddots & \ddots & & & & & & \\
& & & 0 & \lambda & -\lambda & & & & & \\
& & & & & \ddots & \lambda & -\mu & & & \\
\vdots & & & & & & 0\,\lambda+\mu & -\mu & & & \\
& & & & & & & -\lambda & \lambda+\mu & -\mu & \\
0 & & & & & & & & \ddots & \ddots & \ddots \\
-\lambda & & & & & & & & & -\lambda\,\lambda+\mu & -\mu \\
0 & & & & & & & & & & \ddots & \ddots & \ddots \\
\vdots & & & & & & & & & & -\lambda\,\lambda+\mu & -\mu \\
0 & & & & & & & & & & -\lambda & \mu
\end{pmatrix}
\begin{matrix}
1 \\ 2 \\ \vdots \\ \vdots \\ \vdots \\ S-s+1 \\ S-s+2 \\ \vdots \\ 2(S-s) \\ \vdots \\ \vdots \\ m+2S-s
\end{matrix}
$$

The generator matrix A is irreducible, has zero column sum, positive diagonal entries and non-positive off-diagonal entries. Therefore A has one dimensional null space.

8.6 Steady-State Distribution of Inventory Levels

In this section, we derive the analytical form of the steady-state probability distribution of the inventory levels. To obtain the probability distribution vector $\mathbf{p}$, we perform the following row operations on A. Beginning from the last row of A, we add the ith row of A to the $(i-1)$th row of A, for $i = (m+2S-s), (m+2S-s-1), \ldots, 2$. The purpose of these row operations is to eliminate the upper sub-diagonal of A. We get a lower triangular matrix B:

$$
B = \begin{pmatrix}
0 & 0 & & & & & & & & 0 \\
-\lambda & \lambda & 0 & & & & & & & \\
-\lambda & 0 & \ddots & \ddots & & & & & & \\
-\lambda & & \ddots & \lambda & 0 & & & & & \\
-\lambda & & & 0 & \lambda & 0 & & & & \\
-\lambda & & & & 0 & \mu & 0 & & & \\
-\lambda & & & & & -\lambda & \mu & 0 & & \\
\vdots & & & & & & \ddots & \ddots & \ddots & \\
-\lambda & & & & & & & -\lambda & \mu & 0 \\
0 & & & & & & & & \ddots & \ddots & \ddots \\
\vdots & & & & & & & & & -\lambda & \mu & 0 \\
0 & & & & & & & & & & -\lambda & \mu
\end{pmatrix} . \tag{8.15}
$$

By applying the forward substitution to (8.15), we have the following proposition concerning the steady-state probability distribution of the machine-inventory system.

Proposition 8.6.1. *The steady-state probability distribution* $\mathbf{p}$ *of* A *can be written as follows:*

$$\begin{cases} \mathbf{p} = (p_1, p_2, \ldots, p_{m+2S-s}) = \alpha^{-1}(\mathbf{1}, \mathbf{v}, \mathbf{u})^t, \\ \mathbf{1}_{1 \times (S-s)} = (1, 1, \ldots, 1), \\ \rho = \dfrac{\lambda}{\mu}. \end{cases} \quad (8.16)$$

For $\rho \neq 1$, *we have*

$$\begin{cases} \mathbf{v} = \dfrac{\rho}{1-\rho}(1-\rho, \ldots, 1-\rho^{S-s}), \\ \mathbf{u} = \dfrac{\rho(1-\rho^{S-s})}{1-\rho}(\rho, \rho^2, \ldots, \rho^{m+s}), \\ \alpha = \dfrac{\rho^{m+2}(\rho^S - \rho^s) + (1-\rho)(S-s)}{(1-\rho)^2}. \end{cases} \quad (8.17)$$

For $\rho = 1$, *we have*

$$\begin{cases} \mathbf{v} = (1, 2, 3, \ldots, S-s), \\ \mathbf{u} = (S-s, S-s, \ldots, S-s), \\ \alpha = \dfrac{1}{2}(S-s)(3 + S + s + 2m). \end{cases} \quad (8.18)$$

Proof. We first note that if $\mathbf{p}$ is a right null vector of A (i.e. $A\mathbf{p} = 0$) then $\mathbf{p}$ is also a right null vector of B (i.e. $B\mathbf{p} = 0$). We discuss the case when $\rho \neq 1$, the case $\rho = 1$ being similar. Leting $p_1 = \alpha^{-1} > 0$, we observe from (8.15) that $p_i = \alpha^{-1}$ for $i = 1, \ldots, (S-s)$. By forward substitution, we obtain

$$p_{S-s+i} = \frac{\alpha^{-1}\rho}{1-\rho}(1-\rho^i), \quad i = 1, \ldots, (S-s), \quad (8.19)$$

and

$$p_{2(S-s)+j} = \frac{\alpha^{-1}\rho(1-\rho^{S-s})}{1-\rho}\rho^j, \quad j = 1, \ldots, (m+s). \quad (8.20)$$

The normalization constant α obtained by the relation

$$1 = \sum_{i=1}^{m+2S-s} p_i = \alpha^{-1}\left(\frac{\rho^{m+2}(\rho^S - \rho^s) + (1-\rho)(S-s)}{(1-\rho)}\right). \quad (8.21)$$

$\square$

Proposition 8.6.2. *The marginal probability distribution for the inventory levels is*

$$\mathbf{q} = (q_{-m}, \ldots, q_0, q_1, \ldots, q_S)^t. \tag{8.22}$$

Here q_i is the steady-state probability that the inventory level is i, and $i = -m, \ldots, s, \ldots, S$.

For $\rho \neq 1$, we have

$$\alpha = \frac{\rho^{m+2}(\rho^S - \rho^s) + (1-\rho)(S-s)}{(1-\rho)^2},$$

and

$$q_i = \begin{cases} \alpha^{-1} \frac{\rho}{1-\rho}(1-\rho^{S-s})\rho^{r-i}, & i = -m, \ldots, (s-1), s, \\ \alpha^{-1}(1 + \frac{\rho}{1-\rho}(1-\rho^{S-i})), & i = (s+1), \ldots, S. \end{cases} \tag{8.23}$$

For $\rho = 1$, we have $\alpha = \frac{1}{2}(S-s)(3+S+s+2m)$, and

$$q_i = \begin{cases} \alpha^{-1}(S-s), & i = -m, \ldots, 0, \ldots, (s-1), s, \\ \alpha^{-1}(1+S-i), & i = (s+1), \ldots, S. \end{cases} \tag{8.24}$$

Proof. From the ordering of the states (see Fig. 8.1), we have $q_S = p_{S-s}$,

$$q_{S-i} = p_{S-s-i} + p_{S-s+i}, \quad i = 1, 2, \ldots, (S-s-1). \tag{8.25}$$

and

$$q_{-m+i} = p_{m+2S-s-i}, \quad i = 0, 1, \ldots, (m+s). \tag{8.26}$$

The proof is completed by using Proposition 8.6.1. □

8.7 The Average Profit of the Manufacturing System

The average profit of the system can be written as follows:

$$(P - O) * S(s, S) - I(s, S) - P(s, S), \tag{8.27}$$

where $S(s, S), I(s, S)$, and $P(s, S)$ are the average sales, the average inventory holding cost, and the average demand rejection penalty cost respectively. Here P and O are the unit price and the unit operation cost of the product respectively. In terms of the inventory distribution $\mathbf{q}$, the average sales $S(s, S)$, the average inventory cost $I(s, S)$, and the average demand rejection penalty cost $P(s, S)$ can be written as

$$S(s, S) = \lambda \sum_{i=1}^{S} q_i, \quad I(s, S) = c_I \sum_{i=1}^{S} iq_i, \quad \text{and} \quad P(s, S) = c_P \lambda q_{-m} \tag{8.28}$$

respectively. The average profit of the manufacturing system is given by

$$\sum_{i=1}^{S} [\lambda(P - O) - c_I i]q_i - c_P \lambda q_{-m}. \tag{8.29}$$

8.8 Special Case Analysis

In this section, we consider the case of allowing infinite backlog. In this situation, the average profit is given by $S(s, S) - I(s, S)$. We prove that under some conditions, the optimal (s, S) policy is of hedging point type (i.e. $S = (s+1)$) which is the interesting case in [50, 52, 53, 55].

Proposition 8.8.1. *Under the permission of infinite backlog and zero backlog cost, if*

$$0 < \rho < \frac{\sqrt{5} - 1}{2} \quad \text{and} \quad c_I \geq \frac{\lambda(P - O)(1 - \rho)\rho^2}{1 - \rho - \rho^2} \tag{8.30}$$

then the optimal (s, S) policy is of hedging point type, i.e. $S = (s + 1)$.

Proof. Using (8.23) we have

$$S(s, S) - I(s, S) = \frac{(\lambda(P - O) + \frac{c_I}{(1-\rho)})(1 - \rho^2 f(s, S)) - c_I(S + s)}{1 - \rho^{m+2} f(s, S)}, \tag{8.31}$$

where

$$f(s, S) = \frac{\rho^s - \rho^S}{(1 - \rho)(S - s)}. \tag{8.32}$$

When infinite backlog is allowed ($m \to \infty$), the average profit is given by

$$H(s, S) = \left(\lambda(P - O) + \frac{c_I}{1 - \rho}\right)(1 - \rho^2 f(s, S)) - c_I(S + s). \tag{8.33}$$

To maximize $H(s, S)$, it is equivalent to minimize

$$G(s, S) = \left(\lambda(P - O) + \frac{c_I}{1 - \rho}\right)(\rho^2 f(s, S)) + c_I(S + s). \tag{8.34}$$

Let $t = (S - s) \geq 1$, we write

$$G(s, t) = \left(\lambda(P - O) + \frac{c_I}{1 - \rho}\right)(\rho^2 f(s, s + t)) + c_I(t + 2s). \tag{8.35}$$

In the following, we prove that $G(s, t)$ is an increasing function in t for any given s. Thus for any given s, $G(s, t)$ attains its minimum at $t = 1$. Indeed,

$$\frac{\partial G(s, t)}{\partial t} = \frac{(\lambda(P - O)(1 - \rho) + c_I)\rho^{s+2}}{(1 - \rho)^2} \cdot \frac{(\rho^t - 1 - t\rho^t \log \rho)}{t^2} + c_I$$

$$> \frac{(\lambda(P - O)(1 - \rho) + c_I)\rho^{s+2}}{(1 - \rho)^2} \cdot \frac{(\rho^t - 1)}{t^2} + c_I$$

$$\geq \frac{-(\lambda(P-O)(1-\rho)+c_I)\rho^{s+2}}{(1-\rho)}+c_I$$

$$\geq \frac{-(\lambda(P-O)(1-\rho)+c_I)\rho^2}{(1-\rho)}+c_I$$

$$= \frac{-\lambda(P-O)(1-\rho)\rho^2+(1-\rho-\rho^2)c_I}{(1-\rho)}$$

$$\geq 0.$$

The first inequality comes from fact that $-t\rho^t\log\rho > 0$. The last inequality comes from the condition (8.30). The second inequality comes from the fact that the negative function $(\rho^t - 1)/t^2$ is an increasing function in t and we give the proof as follows:

$$\begin{aligned}
\frac{\rho^t - 1}{t^2} &= \frac{\rho-1}{t^2}(1+\rho^1+\cdots+\rho^{(t-1)}) \\
&= \frac{\rho-1}{t(t+1)}\left(1+\frac{1}{t}\right)(1+\rho^1+\cdots+\rho^{(t-1)}) \\
&= \frac{\rho-1}{t(t+1)}\{(1+\rho+\cdots+\rho^{(t-1)})+\frac{1}{t}(1+\rho+\cdots+\rho^{(t-1)})\} \\
&\leq \frac{\rho-1}{t(t+1)}(1+\rho+\cdots+\rho^t) \\
&\leq \frac{\rho-1}{(t+1)^2}(1+\rho+\cdots+\rho^t) \\
&= \frac{\rho^{t+1}-1}{(t+1)^2}.
\end{aligned}$$

Thus the proposition is proved.   □

As a corollary, when the optimal policy is of hedging point type, the average profit is given by (8.33) with $S = (s+1)$, i.e.

$$H(s, s+1) = \left(\lambda(P-O)+\frac{c_I}{1-\rho}\right)(1-\rho^2 f(s, s+1)) - c_I(2s+1). \quad (8.36)$$

We observe that

$$\frac{d^2 H(s, s+1)}{ds^2} = -\left(\lambda(P-O)+\frac{c_I}{1-\rho}\right)\rho^2(\log\rho)^2\rho^s < 0, \quad (8.37)$$

and therefore $H(s, s+1)$ is a convex function in s. Thus the maximum point is given by the unique solution of

$$\frac{dH(s, s+1)}{ds} = -\left(\lambda(P-O)+\frac{c_I}{1-\rho}\right)\rho^2(\log\rho)\rho^s - 2c_I = 0. \quad (8.38)$$

8.9 Numerical Examples

What follows are some numerical examples. Recall that the demand rate

$$\lambda = \sigma P^{-\tau_1} T^{-\tau_2} \quad \text{(cf. 8.1)}.$$

We let the elasticity constants τ_1, τ_2, σ be $2, 1, 1000$ respectively. The parameters P, O, λ, ρ, and c_I are set to satisfy the condition (8.30) in Proposition 8.8.1. The unit price P is 100 and the unit operating cost O is 20. The guaranteed delivery time T is one day. The parameters T_0 and β of the delivery distribution are set to be 0.1 and 10 respectively. The maximum inventory capacity $I_{\max}$ is 100. The inventory cost c_I per unit of product is 10. Furthermore, we let the reliability factor be 90%, and the penalty cost c_P be 2000. In the table below we give the optimal pair (s^*, S^*) and their corresponding average profit $C(s^*, S^*)$ for different values of $\rho = \lambda/\mu$. We found that although the backlog cost c_B and penalty cost c_P are included, the optimal policy is still of hedging point type.

Table 8.1. The optimal (s^*, S^*) and m for different ρ

ρ	m	(s^*, S^*)	$C(s^*, S^*)$
0.1	32	$(1, 2)$	773.1
0.2	20	$(2, 3)$	766.1
0.3	12	$(3, 4)$	757.8
0.4	9	$(4, 5)$	748.4
0.5	7	$(5, 6)$	737.3

8.10 Summary and Discussion

We have discussed the optimal (s, S) policies for production planning with delivery time guarantees. We established a relation (Proposition 8.4.1) between the desirable maximum allowable backlog m and the guaranteed delivery time T. The analytical form of the average profit was also derived. By varying different possible values of s and S, the optimal (s, S) policy can be obtained. A sufficient condition (Proposition 8.8.1) for the hedging point policy to be optimal was also given. It is interesting to relax the condition for hedging point policy to be optimal and extend the model to the case when there are set-up time and set-up costs for the manufacturing system.

Many other interesting problems remain open. for example, what if the machines are unreliable? Extension to the case of multi-items is also interesting. Furthermore, other distributions can be used to model the delivery time.

Another possible extensions of our model is the following non-linear programming problem. We do not predetermine the unit price P and the guaranteed delivery time T (recall that the demand rate is a function of P and T cf. (8.1)). Rather, they are part of the decisions. The generalized formulation is given as follows:

$$\max_{P,T,s,S} \sum_{i=1}^{S} [\lambda(P - O) - c_I i]q_i - c_P \lambda q_{-m}$$

$$\text{subject to} \begin{cases} (8.11), \\ s < S \le I_{\max}, \\ \text{and } 1 \le m. \end{cases}$$

Exercises

8.1 Suppose the service time of a customer follows the exponential distribution with parameter λ. A delivery time guarantee is given to the customer with a service level of γ. What is the minimum guaranteed time T?

8.2 A repairman planned to attract customers by giving them a repair time guarantee. The repairing process consists of two stages: investigation and fixing. The customers are served in a first-come-first-served manner. From his past experience, the mean investigation time is 10 min and the mean fixing time is 20 min. If he wants to guarantee at least 90% service level, what is the best guaranteed delivery time T?

8.3 Refer to Question 8.2. After the implementation of DTG policy, the repairman still not satisfies with his profit and he becomes more aggressive. Every day, on average he fixes eight computers. To attract more customers, he decided to guarantee each customer a 45 min repairing policy. Customers will get a free service if he cannot finish the repairing within 45 min. He expects the demand will increase to ten. Assuming that his prediction is true, will his expected income be increased?

9. Multi-Location and Two-Echelon Inventory Systems

9.1 Introduction

In this chapter, we are going to study the applications of Markovian queues in *supply chain* problems. The first model is related to an interesting application of the MMPPs in *multi-location inventory systems*. The second model is a *two-echelon inventory system*. The inventory models studied here are *make-to-order* type which is different from those manufacturing systems studied in previous chapters. This means that when queuing network is applied to modeling the inventory system, the waiting spaces in the queue are not inventories to be held but they are the orders to be satisfied.

In Chapter 8, we have discussed the fact that efficiency of product/service delivery (delivery time guarantee policy) is one of the major concerns in many industries. The reliability and timeliness of product/service delivery are powerful weapons for competing in the market. It is common to have several parallel locations to provide service to various local markets [39, 40, 42]. These locations normally replenish their stocks from a common main depot; see for instance Axsäter [7] and Kochel [85]. We consider inventory systems of single consumable items in multi-location situations. In the event of items being out of stock at a location, the order is transferred to the main depot. We model the inventory system of each location and the main depot by Markovian queuing networks. The transshipment can be modeled by the Markov-Modulated Poisson Process (MMPP) (discussed in Chapter 3) which is a generalization of the Poisson process.

In Section 9.3 we consider a two-echelon inventory model. The inventory system consists of an infinite *supply plant*, a *central warehouse*, and a number of *local warehouses*. Since the customers are usually scattered over a large regional area, a network of inventory locations (local warehouses) is necessary to maintain a high service level. In particular, in our study *Emergency Lateral Transshipment* (ELT) is allowed among the local warehouses to enhance the service level. Quite a number of research studies have been done in the area of ELT. We refer readers to the following interesting references. Moinzadeh and Schmidt [91] have studied the emergency replenishments for a *single-echelon model* with deterministic lead times. Aggarwal and Moinzadeh [1] then extended the idea to a two-echelon model. Lee [87] and Axsäter [7] considered a two-echelon system in which the local warehouses are grouped together. Within the group, the warehouses were assumed to be identical. Approximation expressions for the demand satisfied by ELT were obtained. A simulation study on the two-echelon system can also be found in Pyke [97].

The rest of the chapter is organized as follows. In Section 9.2 we discuss the multi-location inventory system. In Section 9.2.1, we discuss the queuing network of each location. In Section 9.2.2, we determine the MMPP parameters of the transshipment process. We then give the generator matrix for the depot queue in Section 9.2.3. In Section 9.2.4, we discuss the solution of the steady-state probability distribution of the depot queue. In Section 9.3, we present the proposed two-echelon inventory model. The assumptions and notations of the model are also given. In Section 9.3.1, we consider a Markovian queuing model for the local warehouses. In Section 9.3.2, we study an aggregated model for the central warehouse. Numerical algorithms are presented to solve the steady-state probability distribution of the system. Finally a summary is given in Section 9.4 to conclude this chapter.

9.2 The Multi-Location Inventory Systems

In the first model, we consider an m-location and one *main depot* inventory system under *continuous review* and *one-for-one replenishments*. The demand for each location and the depot is described by the mutually independent Poisson processes. The service time (or lead time) for each demand is exponentially distributed. In each location, all the unsatisfied demand are backlogged up to a given level and will overflow to the main depot whenever this level is attained. The inputs of the depot are its own demand and the transshipments from all the m parallel locations. A *transshipment cost* is assigned for each unit of overflowed demand. In the depot, all the unsatisfied demand are also backlogged up to a given level and the demand will be rejected if this level is attained. A *rejection cost* is assigned to each unit of rejected demand.

For each location, the inventory system is modeled by the $M/M/s/q$ queue with the following relations:

(a) s, the inventory level (number of servers),
(b) μ^{-1}, the mean lead time (mean service time),
(c) λ^{-1}, the mean inter-arrival time of demand,
(d) q, the number of backlogs (queue length).

The *overflow* (transshipment) process of the demand of each location is modeled by a two-state MMPP. The two states are the queue is full (the maximum level of backlogs is attained) and the queue is not yet full (further demand is acceptable) respectively. The MMPP approximation assumption reduces the complexity and the number of states of the $(m + 1)$ queuing networks; see Meier-Hellstern [90]. The approximation is also shown to be accurate and reasonable; see also [50, 72, 90, 116] and Chapter 3. The duration of the first and the second state are assumed to be exponentially distributed with means σ_1^{-1} and σ_2^{-1} respectively. When the queue is in the first state, it will overflow the demand to the depot at a rate of λ. Thus the generator matrix for two-state Markov chain and its corresponding rate matrix (cf. Chapter 3) are given by

$$Q_\sigma = \begin{pmatrix} \sigma_1 & -\sigma_2 \\ -\sigma_1 & \sigma_2 \end{pmatrix} \quad \text{and} \quad \Lambda_\sigma = \begin{pmatrix} \lambda & 0 \\ 0 & 0 \end{pmatrix}. \tag{9.1}$$

respectively. For the main depot queue, the input comes from its own demand and the overflow of all the locations which is a superposition of m independent two-state MMPPs. Thus the inventory system of the depot is an (MMPP/M/s/q) queue. We will use the following notation in this section:

(i) λ^{-1}, the mean arrival time of demand of the depot,
(ii) μ^{-1}, the mean service time (lead time) of each server of the depot,
(iii) s, the number of servers (inventory) in the depot,
(iv) q, the number of waiting spaces (maximum backlog) in the depot,
(v) m, the number of locations,
(vi) λ_i^{-1}, the mean arrival time of demand of location $i(i = 1, 2, \ldots, m)$,
(vii) μ_i^{-1}, the mean service time (lead time) of each server of location $i(i = 1, 2, \ldots, m)$,
(viii) s_i, the number of servers (inventory) in location $i(i = 1, 2, \ldots, m)$,
(ix) m_i, the number of waiting spaces (maximum backlog) in location $i(i = 1, 2, \ldots, m)$,
(x) (Q_i, Λ_i), $1 \le i \le m$, the parameters of the MMPP's modeling overflow demand, where

$$Q_i = \begin{pmatrix} \sigma_{i1} & -\sigma_{i2} \\ -\sigma_{i1} & \sigma_{i2} \end{pmatrix} \quad \text{and} \quad \Lambda_i = \begin{pmatrix} \lambda_i & 0 \\ 0 & 0 \end{pmatrix}, \tag{9.2}$$

(xi) h_i, the unit inventory cost in location $i(i = 1, 2, \ldots, m)$,
(xii) w_i, the unit waiting cost in location $i(i = 1, 2, \ldots, m)$,
(xiii) τ_i, the unit transshipment cost in location $i(i = 1, 2, \ldots, m)$,
(xiv) h, the unit inventory cost in the depot,
(xv) w, the unit waiting cost in the depot,
(xvi) p, the profit per satisfied demand,
(xvii) r, the rejection cost of an unit of demand.

9.2.1 The Queuing System of Each Location

In location i, the inventory system can be modeled by the M/M/s/q queue. Using the well-known result, the steady-state probability that location i has an inventory level of j is given by

$$p_i(j) = \alpha_i^{-1} \left(\prod_{k=1}^{s_i-j} \frac{\lambda_i}{\mu_i \min\{k, s_i\}} \right), \quad j = -q_i, \ldots, 0, \ldots, s_i, \tag{9.3}$$

where

$$\alpha_i = \sum_{j=-q_i}^{s_i} \left(\prod_{k=1}^{s_i-j} \frac{\lambda_i}{\mu_i \min\{k, s_i\}} \right). \tag{9.4}$$

Here negative inventory means backlog. By using the above result, we have the following proposition.

Proposition 9.2.1. *In location i the throughput is given by*

$$T_i = \sum_{k=-q_i}^{s_i} p_i(k)\mu_i \min\{s_i - k, s_i\}, \tag{9.5}$$

and the average profit is given by

$$g_i(s_i, q_i, \mu_i, \lambda_i) = pT_i - h_i \sum_{k=1}^{s_i}(s_i - k)p_i(k) - w_i \sum_{k=1}^{q_i} kp_i(-k) - \tau_i \lambda_i p_i(-q_i). \tag{9.6}$$

Proof. We note that the expected gain is pT_i, the expected holding cost is

$$h_i \sum_{k=1}^{s_i}(s_i - k)p_i(k), \tag{9.7}$$

the expected backlog waiting cost is

$$w_i \sum_{k=1}^{q_i} kp_i(-k) \tag{9.8}$$

and the expected transshipment cost is

$$\tau_i \lambda_i p_i(-q_i). \tag{9.9}$$

Thus we have the average profit of location i given by

$$pT_i - h_i \sum_{k=1}^{s_i}(s_i - k)p_i(k) - w_i \sum_{k=1}^{q_i} kp_i(-k) - \tau_i \lambda_i p_i(-q_i). \tag{9.10}$$

$\square$

9.2.2 Determination of the MMPP Parameters

The overflow of demand (transshipment) is modeled by a two-state Markov-Modulated Poisson Process (MMPP) in each location i. The two states are the queue is full (the maximum level of backlogs is attained) and the queue is not yet full (further demand is acceptable) respectively. The duration of the first and second state are assumed to be exponentially distributed with means σ_{i1}^{-1} and σ_{i2}^{-1} respectively. Here we determine the values of σ_{i1} and σ_{i2} under the MMPP assumption. We derive the probability density function of the duration time of the first state. We observe that in location i when the queue is full (first state), the s_i parallel servers are busy. Therefore the transition of state occurs when any one of the busy servers finishes a demand. If $\{t_1, \ldots, t_{s_i}\}$ are times for the servers to finish the demand on hand then the change of state occurs at $\min\{t_1, \ldots, t_{s_i}\}$. The following lemma gives the probability density function of $\min\{t_1, \ldots, t_{s_i}\}$.

Lemma 9.2.1. *Let* $t_1, \ldots, t_{s_i}$ *be* s_i *independent random variables follow the exponential distribution*

$$\mu_i e^{-\mu_i x}, \quad 0 \leq x,$$

and $t = \min\{t_1, \ldots, t_{s_i}\}$. *Then the variable* t *is still exponentially distributed and has mean* $(s_i \mu_i)^{-1}$.

Proposition 9.2.2. *The MMPP parameters for location* $i(i = 1, 2, \ldots, m)$ *are given by*

$$\sigma_{i1} = s_i \mu_i \quad \text{and} \quad \sigma_{i2} = \frac{b_i}{1 - b_i} s_i \mu_i \ (i = 1, 2, \ldots, m),$$

where b_i *is the blocking probability of location* i *cf. (9.3).*

Proof. From Lemma 9.2.1, we have $\sigma_{i1} = s_i \mu_i$. From (9.2) the steady-state probability of the two-state MMPP is given by the vector

$$\left(\frac{\sigma_{i2}}{\sigma_{i1} + \sigma_{i2}}, \frac{\sigma_{i1}}{\sigma_{i1} + \sigma_{i2}} \right)^t. \tag{9.11}$$

Let $b_i = p_i(s_i)$ (cf. (9.3)) be the blocking probability of the location i. Since the first state of the MMPP corresponds to the situation that the queue is blocked, we have the following ratio:

$$\frac{\text{Time of blocking}}{\text{Time of no blocking}} = \frac{b_i}{1 - b_i} = \frac{\sigma_{i2}}{\sigma_{i1}}. \tag{9.12}$$

Thus we have

$$\sigma_{i2} = \frac{b_i}{1 - b_i} s_i \mu_i. \tag{9.13}$$

$\square$

9.2.3 The Queuing Systems of the Depot

In this section, we give the generator matrix for the depot queue. The input of the main depot queue comes from its own demand with rate λ and the superposition of all the overflows from the m locations. The superposition of several independent MMPPs, which is still an MMPP and is parameterized by two $2^m \times 2^m$ matrices (Q, Γ). Here

$$Q = (Q_1 \otimes I_2 \otimes \cdots \otimes I_2) + (I_2 \otimes Q_2 \otimes I_2 \otimes \cdots \otimes I_2) + \cdots + (I_2 \otimes \cdots \otimes I_2 \otimes Q_m),$$
$$(9.14)$$
$$\Lambda = (\Lambda_1 \otimes I_2 \otimes \cdots \otimes I_2) + (I_2 \otimes \Lambda_2 \otimes I_2 \otimes \cdots \otimes I_2) + \cdots + (I_2 \otimes \cdots \otimes I_2 \otimes \Lambda_m) \quad (9.15)$$

and

$$\Gamma = \Lambda + \lambda I_{2^m}, \tag{9.16}$$

where I_2 and I_{2^q} are the 2×2 and $2^m \times 2^m$ identity matrices respectively. We can regard our (MMPP/M/s/q) queue as a Markov process on the state space

$$\{(i, j) \mid -q \le i \le s, 1 \le j \le 2^m\}.$$

The number i corresponds to the inventory level at the depot, while j corresponds to the state of the Markov process with generator matrix Q. Hence the generator matrix of the queuing process is given by the following $(s + q + 1)2^m \times (s + q + 1)2^m$ tri-diagonal block matrix A:

$$A = \begin{pmatrix} Q + \Gamma & -\mu I & & & & 0 \\ -\Gamma & Q + \Gamma + \mu I & -2\mu I & & & \\ & \ddots & \ddots & \ddots & & \\ & & -\Gamma & Q + \Gamma + s\mu I & -s\mu I & \\ & & & \ddots & \ddots & \ddots \\ & & & & -\Gamma & Q + \Gamma + s\mu I & -s\mu I \\ 0 & & & & & -\Gamma & Q + s\mu I \end{pmatrix}.$$
$$(9.17)$$

From the steady-state probability distribution of G, we have the following proposition.

Proposition 9.2.3. *For the main depot, the throughput is given by*

$$T = \sum_{k=-q}^{s} r(k)\mu \min\{s - k, s\},$$

and the average profit is

$$g(s, q, \mu, \lambda) = pT - h \left(\sum_{k=1}^{s} (s - k) r(k) \right) - w \left(\sum_{k=1}^{q_i} k r(k) \right)$$
$$- r(1, 1, \ldots, 1) \Gamma(p(-q, 1), \ldots, p(-q, 2^m))^t.$$

Here

$$r(k) = \sum_{j=1}^{2^m} p(k, j)$$

is the marginal steady-state probability of the inventory levels.

Proof. We note that the expected gain is pT, the expected holding cost is

$$h \sum_{k=1}^{s} (s - k)r(k),$$

the expected backlog waiting cost is

$$w \sum_{k=1}^{q} kr(-k)$$

and the expected rejection cost is

$$r(1, 1, \ldots, 1)\Gamma(p(-q, 1), \ldots, p(-q, 2^m))^t.$$

Thus we have the average profit given by

$$pT - h \left(\sum_{k=1}^{s} (s - k)r(k) \right) - w \left(\sum_{k=1}^{q} kr(k) \right)$$
$$-r(1, 1, \ldots, 1)\Gamma(p(-q, 1), \ldots, p(-q, 2^m))^t.$$

$\square$

9.2.4 The Steady-State Probability of the Depot Queue

Usually a classical iterative method such as the block Gauss-Seidel is used to solve the steady-state probability vector ($A\mathbf{p} = \mathbf{0}$ and $\mathbf{1}^t\mathbf{p} = 1$). In general this method has a slow convergence rate. For general values of s, the Preconditioned Conjugate Gradient (PCG) method discussed in Chapter 3 can solve the vector $\mathbf{p}$ very efficiently. In the remainder of this section, an asymptotic solution is derived for the steady-state probability $\mathbf{p}$ when m is large under certain reasonable conditions. We also give a formulation for the multi-location inventory problem.

We recall that the size of the matrix G is $2^m(s + q + 1) \times 2^m(s + q + 1)$. When m (the number of locations) increases, the cost of solving the steady-state probability vector increases exponentially. Therefore it is natural to seek an analytical approximation for this vector. In the following, we propose an approximation $\mathbf{v}$ such that

$$\lim_{m \to \infty} ||A\mathbf{v}||_\infty = 0$$

under the assumption

$$0 < \delta < b_i \le 0.5 \quad (i = 1, 2, \ldots, m).$$

The assumption implies that the blocking probabilities of all the locations are greater than a positive δ and smaller than 0.5, which is reasonable in general. We observe that the generator matrix A has the following tensor structure:

$$A = I_n \otimes Q + C \otimes \Lambda + T \otimes I_{2^m} \tag{9.18}$$

where

$$T = \begin{pmatrix} \lambda & -\mu & & & & & 0 \\ -\lambda & \lambda + \mu & -2\mu & & & & \\ & \ddots & \ddots & \ddots & & & \\ & & -\lambda & \lambda + s\mu & -s\mu & & \\ & & & \ddots & \ddots & \ddots & \\ & & & & -\lambda & \lambda + s\mu & -s\mu \\ 0 & & & & & -\lambda & s\mu \end{pmatrix}, \tag{9.19}$$

$$C = \begin{pmatrix} 1 & & & & 0 \\ -1 & 1 & & & \\ & \ddots & \ddots & & \\ & & -1 & 1 & \\ 0 & & & -1 & 0 \end{pmatrix}, \tag{9.20}$$

and $n = s + q + 1$.

Proposition 9.2.4. *Let* **x** *and* **y** *be the steady-state probability distributions for the generator matrices T and Q respectively. If*

$$0 < \delta < b_i \le 0.5 \quad (i = 1, 2, \ldots, m),$$

for a positive δ and all λ_i are bounded and independent of m then

$$\lim_{m \to \infty} \|A(\mathbf{x} \otimes \mathbf{y})\|_\infty = 0.$$

The probability vector

$$\mathbf{v} = \mathbf{x} \otimes \mathbf{y}$$

is an asymptotic solution to the generator A for large m.

Proof. The probability vector **x** is just the solution of (9.3) with

$$s_i = s, \quad q_i = q, \quad \lambda_i = \lambda \quad \text{and} \quad \mu_i = \mu.$$

By direct verification, we have the solution of

$$\Lambda \mathbf{y} = \mathbf{0}$$

given by

$$\mathbf{y} = \left(\frac{\sigma_{12}}{\sigma_{11} + \sigma_{12}}, \frac{\sigma_{11}}{\sigma_{11} + \sigma_{12}} \right)^t \otimes \left(\frac{\sigma_{22}}{\sigma_{21} + \sigma_{22}}, \frac{\sigma_{21}}{\sigma_{21} + \sigma_{22}} \right)^t$$

$$\otimes \cdots \otimes \left(\frac{\sigma_{m2}}{\sigma_{m1} + \sigma_{m2}}, \frac{\sigma_{m1}}{\sigma_{m1} + \sigma_{m2}} \right)^t$$

From (9.18) we get

$$
\begin{aligned}
A\mathbf{v} &= A(\mathbf{x} \otimes \mathbf{y}) \\
&= (I_n \otimes Q + C \otimes \Lambda + T \otimes I_{2^m})(\mathbf{x} \otimes \mathbf{y}) \\
&= (C\mathbf{x}) \otimes (\Lambda\mathbf{y}).
\end{aligned}
$$

We observe

$$||C||_\infty = 2, \quad ||\mathbf{x}||_\infty \le 1$$

and

$$||\Lambda||_\infty = m \max_i \{\lambda_i\}.$$

The $||H||_\infty$ of a $p \times q$ matrix is given as follows:

$$||H||_\infty = \max_i \left\{ \sum_{j=1}^{q} |H_{ij}| \right\}.$$

From Proposition 9.2.2, we have

$$\frac{\sigma_{i1}}{\sigma_{i1} + \sigma_{i2}} = 1 - b_i \le 1 - \delta \quad \text{and} \quad \frac{\sigma_{i2}}{\sigma_{i1} + \sigma_{i2}} = b_i.$$

By the assumption that $b_i \le 0.5$, we have $1 - b_i \ge b_i$. Thus we have

$$||\mathbf{y}||_\infty \le (1 - \delta)^m.$$

Therefore we have

$$\lim_{m \to \infty} ||A(\mathbf{x} \otimes \mathbf{y})||_\infty \le \lim_{m \to \infty} 2m(\max_i \{\lambda_i\})(1 - \delta)^m = 0.$$

$\square$

We end this section by giving the formulation of the multi-location inventory problem as follows. Given $\lambda, \lambda_i, \mu, \mu_i, q$, and q_i, one interesting optimization problem is the following:

$$\max_{s, s_i} \left\{ g(s, q, \mu, \lambda) + \sum_{i=1}^{m} g_i(s_i, q_i, \mu_i, \lambda_i) \right\} \quad \text{subject to} \quad s + \sum_{i=1}^{m} s_i \le \tilde{S}. \tag{9.21}$$

9.3 The Two-Echelon Inventory Model

In this section we study an analytical two-echelon inventory model in a *supply chain problem*. The model consists of an infinite supply plant, a central warehouse, and n identical local warehouses; see Fig. 9.1. The demand at each local warehouse is assumed to follow a Poisson process with mean arrival rate λ and is served in the following manner:

(a) The demand is first fulfilled by the stock at the local warehouse and at the same time a replenishment order is issued to the central warehouse. The replenishment time from the central warehouse to a local warehouse is assumed to be exponentially distributed with mean μ^{-1}. We remark that the demand is served in a "first-come-first-served" principle and the replenishment is carried out in a "one-for-one" manner.

(b) If the local warehouse is out of stock, the demand is satisfied by an ELT from another local warehouse (randomly); at the same time the local warehouse that sources the ELT will issue a replenishment order to the central warehouse. Here in this situation, we assume that the ELT time is shorter than a replenishment order from the central warehouse.

(c) If, unfortunately, all the other local warehouses are also out of stock (the demand cannot be satisfied by an ELT), then the demand is satisfied by a direct delivery from the central warehouse and at the same time a replenishment order is issued to the plant from the central warehouse. The replenishment time from the plant to the central warehouse is assumed to be exponentially distributed with mean γ^{-1}. In this case, we assume that the time for the direct delivery from the central warehouse is shorter than the waiting time for a normal replenishment from the central warehouse.

(d) Finally, if all the warehouses (including the central warehouse) are out of stock, the demand is then satisfied by direct delivery from the plant.

There are costs associated with the normal replenishments and ELT. Moreover there are inventory costs at each local warehouse and the central warehouse. We are going to study the system in three stages. We first analyze the inventory in each local warehouse as a simple Markovian queuing system. Secondly, we consider an aggregated model for the inventory levels in the central warehouse and the local warehouses. Finally, we integrate the results in the first two stages to obtain an average running cost. The following notation will be used for our discussion throughout the rest of the chapter.

(i) n, the number of local warehouses,
(ii) L, the maximum inventory level in each local warehouse,
(iii) C, the maximum inventory level in the central warehouse,
(iv) λ^{-1}, the mean waiting time of a demand in each local warehouse,
(v) μ^{-1}, the mean waiting time of a replenishment from the central warehouse to a local warehouse,

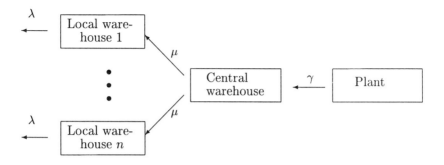

Fig. 9.1. The two-echelon inventory system.

(vi) γ^{-1}, the mean waiting time of a replenishment from the plant to the central warehouse,

(vii) I_L, the unit inventory cost of a local warehouse,

(viii) I_C, the unit inventory cost of the central warehouse,

(ix) R_C, the unit replenishment cost from the central warehouse to a local warehouse,

(x) R_P, the unit replenishment cost from the plant to the warehouse,

(xi) E_L, the unit emergency lateral transshipment cost,

(xii) D_C, the unit emergency delivery cost from the central warehouse to the customers when all the local warehouses including the central warehouse are out of stock,

(xiii) D_P, the unit emergency delivery cost from the plant to the customers when all the warehouses are out of stock.

9.3.1 The Local Warehouse

Since the maximum inventory in a local warehouse is L, the possible inventory levels take integer values in the set $\{0, 1, \ldots, L\}$; see Fig. 9.2. We note that the demand consists of the exogenous demand rate λ and also the ELT from the other $n-1$ warehouses. Suppose that the "starving (out of stock) probability" of each local warehouse is p, then the rate of ELT from the $n-1$ warehouses will be $(n-1)p\lambda$. Thus the demand rate of a local warehouse is given by

$$\tau = \lambda + (n-1)p\lambda. \tag{9.22}$$

According to the flow of stock (orders) described in (a),(b),(c) and (d) of Section 9.3, the replenishment rate is $(L-i)\mu$ when the inventory level is i. We let

$$\mathbf{p} = (p_0, p_1, \ldots, p_L)^t$$

be the steady-state probability vector for the local warehouse system. Here p_i is the steady-state probability that the inventory level is i. We order the

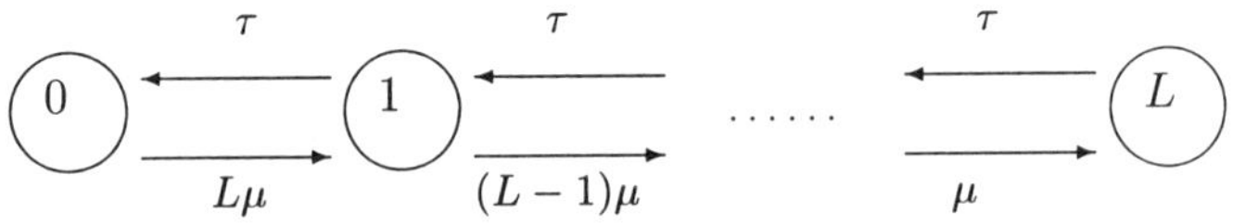

Fig. 9.2. The Markov chain of the local warehouse.

states from 0 to L and obtain the following generator matrix A for the system and

$$
A = \begin{array}{c} 0 \\ 1 \\ 2 \\ 3 \\ \vdots \\ L \end{array}
\left(\begin{array}{ccccccc}
* & -\tau & & & & & 0 \\
-L\mu & * & -\tau & & & & \\
& -(L-1)\mu & * & -\tau & & \\
& & & \ddots & \ddots & \ddots & \\
& & & & -2\mu & * & -\tau \\
0 & & & & & -\mu & *
\end{array} \right). \tag{9.23}
$$

Here "$*$" is defined such that each column sum equals zero. The steady-state probability vector $\mathbf{p}$ satisfies

$$
A\mathbf{p} = \mathbf{0} \quad \text{and} \quad \sum_{i=0}^{L} p_i = 1. \tag{9.24}
$$

It is well known that

$$
p_i = \frac{(\tau/\mu)^{L-i}}{(L-i)!} \left(\sum_{j=0}^{L} \frac{(\tau/\mu)^j}{j!} \right)^{-1}. \tag{9.25}
$$

Since p_0 is the probability that the local warehouse is out of stock, from (9.22) and (9.25) we have

$$
\begin{aligned}
p &= p_0 \\
&= \frac{(\tau/\mu)^L}{(L)!} \left(\sum_{j=0}^{L} \frac{(\tau/\mu)^j}{j!} \right) \\
&= \frac{(\lambda(1+(n-1)p)/\mu)^L}{(L)!} \left(\sum_{j=0}^{L} \frac{(\lambda(1+(n-1)p)/\mu)^j}{j!} \right)^{-1}.
\end{aligned} \tag{9.26}
$$

To study the starving probability p, we consider the following function:

$$f(p) = \frac{(\lambda(1 + (n-1)p)/\mu)^L}{(L)!} - p\left(\sum_{j=0}^{L} \frac{(\lambda(1 + (n-1)p)/\mu)^j}{j!}\right). \qquad (9.27)$$

We note that the starving probability p of a local warehouse satisfies $f(p) = 0$ and $0 \leq p \leq 1$. We are going to show that $f(p) = 0$ has exactly one zero in $(0, 1)$ under some conditions. We need the following lemma.

Lemma 9.3.1. *For $L = 1, 2, \ldots$, we have*

$$\sum_{j=0}^{L} \frac{\alpha^j}{j!} - \frac{L\alpha^{L-1}}{(L-1)!} \geq 0 \quad \text{for} \quad 0 \leq \alpha \leq 1.$$

Proof. We note that

$$\frac{\alpha^j}{j!} \geq \frac{\alpha^{L-1}}{(L-1)!} \quad \text{for } 0 \leq \alpha \leq 1, \quad \text{and } j = 0, 1, 2, \ldots, L-1.$$

By taking the summation over j we have

$$\sum_{j=0}^{L-1} \frac{\alpha^j}{j!} \geq \frac{L\alpha^{L-1}}{(L-1)!} \quad \text{for} \quad 0 \leq \alpha \leq 1.$$

Hence the result is proved by adding the positive term $\frac{\alpha^L}{L!}$ in the summation. $\qquad \square$

Proposition 9.3.1. *If $n\lambda \leq \mu$ then*

$$f(p) = \frac{(\lambda(1 + (n-1)p)/\mu)^L}{(L)!} - p\left(\sum_{j=0}^{L} \frac{(\lambda(1 + (n-1)p)/\mu)^j}{j!}\right) = 0$$

has an unique zero in $(0, 1)$. The root p is the common starving probability in each location. The condition $n\lambda \leq \mu$ simply means that the demand rate of all the local warehouses $(n\lambda)$ is not greater than the replenishment rate μ.

Proof. Since

$$f(0) = \frac{(\lambda/\mu)^L}{(L)!} > 0,$$

$$f(1) = -\left(\sum_{j=0}^{L-1} \frac{(n\lambda/\mu)^j}{j!}\right) < 0$$

and $f(p)$ is continuous, $f(p)$ has at least one zero in $(0, 1)$. We note that

$$f'(p) = \frac{(\lambda(1 + (n-1)p)/\mu)^{L-1}}{(L-1)!} \frac{\lambda(n-1)}{\mu} - \sum_{j=0}^{L} \frac{(\lambda(1 + (n-1)p)/\mu)^j}{j!}$$

$$- \frac{p\lambda(n-1)}{\mu} \left(\sum_{j=0}^{L-1} \frac{(\lambda(1 + (n-1)p)/\mu)^j}{j!} \right). \tag{9.28}$$

From the condition $n\lambda \le \mu$, we have

$$\frac{\lambda}{\mu}(1 + (n-1)p) \le \frac{n\lambda}{\mu} \le 1 \quad \text{and} \quad \lambda(n-1)/\mu \le 1.$$

Let

$$\alpha = \frac{\lambda}{\mu}(1 + (n-1)p),$$

the first two terms in (9.28) is non-positive by Lemma 9.3.1. Therefore we have $f'(p) < 0$ and this implies that $f(p)$ is strictly decreasing in $(0, 1)$ and hence $f(p) = 0$ has exactly one root in $(0, 1)$. $\qquad\Box$

Unfortunately there is no analytical solution for the starving probability p; however, one may apply the numerical method, namely the bisection method to obtain an approximated solution. We remark that in the tested numerical examples, even though the condition in Proposition 9.3.1 is not satisfied, a unique zero of $f(p)$ is found in $(0, 1)$. By using the bisection method, we compute the starving probabilities for different values of L, n and λ/μ in the following table. We observe that the starving probability is very sensitive to L (the maximum inventory in the local warehouse) and decreases very fast when L increases. However, for large value of L, the starving probability is not very sensitive to the value of n (number of local warehouses).

From (9.25) and above, the expected inventory cost, expected normal replenishment cost and expected ELT cost of the local warehouses are given respectively by

$$\begin{cases} nI_L \sum_{i=0}^{L} ip_i, \\ nR_C \sum_{i=0}^{L} (L-i)p_i, \\ nE_L\lambda p. \end{cases} \tag{9.29}$$

9.3.2 An Aggregated Model for Central Warehouse

In this section, we discuss an aggregated inventory model for the central warehouse and the local warehouses by aggregating the demand and inventory of the local warehouses; see Fig. 9.3. The aggregated demand for the central warehouse is the superposition of all the demand from the local warehouses which is still a Poisson process. In the model, we let $x(t)$ be the inventory at the central warehouse and $y(t)$ be the total inventory level of the sum of all the local warehouse at time t.

Table 9.1. The starving probability p of the system for different λ/μ

$\lambda/\mu = 0.2$	$L = 2$	$L = 4$	$L = 6$	$L = 8$	$L = 10$
$n = 2$	0.0169	5.46×10^{-5}	7.28×10^{-8}	5.2×10^{-11}	2.31×10^{-14}
$n = 4$	0.0180	5.46×10^{-5}	7.28×10^{-8}	5.2×10^{-11}	2.31×10^{-14}
$n = 6$	0.0193	5.46×10^{-5}	7.28×10^{-8}	5.2×10^{-11}	2.31×10^{-14}
$n = 8$	0.0209	5.46×10^{-5}	7.28×10^{-8}	5.2×10^{-11}	2.31×10^{-14}
$\lambda/\mu = 0.5$	$L = 2$	$L = 4$	$L = 6$	$L = 8$	$L = 10$
$n = 2$	0.0874	0.0016	1.32×10^{-5}	5.88×10^{-8}	1.63×10^{-10}
$n = 4$	0.1216	0.0016	1.32×10^{-5}	5.88×10^{-8}	1.63×10^{-10}
$n = 6$	0.2000	0.0016	1.32×10^{-5}	5.88×10^{-8}	1.63×10^{-10}
$n = 8$	0.3585	0.0016	1.32×10^{-5}	5.88×10^{-8}	1.63×10^{-10}
$\lambda/\mu = 1$	$L = 2$	$L = 4$	$L = 6$	$L = 8$	$L = 10$
$n = 2$	0.2600	0.0161	5.12×10^{-4}	9.12×10^{-6}	1.01×10^{-7}
$n = 4$	0.4531	0.0180	5.15×10^{-4}	9.13×10^{-6}	1.01×10^{-7}
$n = 6$	0.6236	0.0206	5.18×10^{-4}	9.13×10^{-6}	1.01×10^{-7}
$n = 8$	0.7812	0.0245	5.20×10^{-4}	9.13×10^{-6}	1.01×10^{-7}

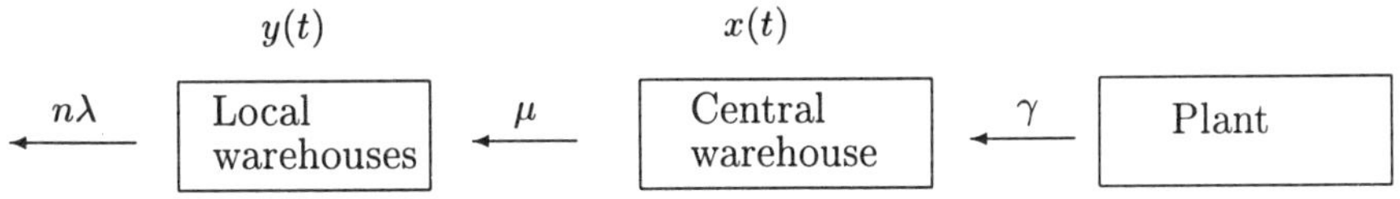

Fig. 9.3. The aggregated model.

According to the flow of stock (orders) described in (a),(b),(c) and (d) in Section 9.3, the states of the joint inventory process

$$\{(x(t), y(t)), t \geq 0\} \tag{9.30}$$

form a continuous time Markov chain taking values in the state space

$$S = \{(x(t), y(t)) : x(t) = 0, \dots, C; \; y(t) = 0, \dots, nL\}. \tag{9.31}$$

The total number of states is $(C+1)(nL+1)$. Each time when visiting a state, the process stays there for a random period of time that has an exponential distribution and is independent of the past behavior of the process. If we order the inventory level at the central warehouse and the inventory level of the local warehouses lexicographically (the inventory levels are ordered in increasing order), the steady-state probability distribution $\mathbf{q}$ is the solution of the following linear system:

$$Bq = 0 \quad \text{and} \quad \sum_{i=1}^{(C+1)(nL+1)} q_i = 1 \tag{9.32}$$

where B is the generator matrix for the inventory system which we will describe later and the column vector q is the following vector:

$$(q_1, \ldots, q_{nL+1}, \ldots, q_{nL+2}, \ldots, q_{2*(nL+1)}, \ldots, q_{C(nL+1)+1}, \ldots, q_{(C+1)(nL+1)})^t$$

which is the required steady-state probability distribution. Here $q_{i*(nL+1)+j}$ is the steady-state probability that the inventory level at the central warehouse is i and the total inventory level of the local warehouses is j. In the following let us give a simple example.

Example 9.3.1. When $C = 2$, $L = 1$, and $n = 3$. The generator matrix is given by the following 12×12 matrix:

$$
\begin{array}{c}
(0,0) \\ (0,1) \\ (0,2) \\ (0,3) \\ (1,0) \\ (1,1) \\ (1,2) \\ (1,3) \\ (2,0) \\ (2,1) \\ (2,2) \\ (2,3)
\end{array}
\left(
\begin{array}{cccccccccccc}
* & -3\lambda & 0 & 0 & -3\lambda & -3\lambda & 0 & 0 & 0 & 0 & 0 & 0 \\
-3\mu & * & -3\lambda & 0 & 0 & 0 & -3\lambda & 0 & 0 & 0 & 0 & 0 \\
0 & -2\mu & * & -3\lambda & 0 & 0 & 0 & -3\lambda & 0 & 0 & 0 & 0 \\
0 & 0 & -\mu & * & 0 & 0 & 0 & 0 & 0 & 0 & 0 & 0 \\
-2\gamma & 0 & 0 & 0 & * & 0 & 0 & 0 & -3\lambda & -3\lambda & 0 & 0 \\
0 & -2\gamma & 0 & 0 & -3\mu & * & 0 & 0 & 0 & 0 & -3\lambda & 0 \\
0 & 0 & -2\gamma & 0 & 0 & -2\mu & * & 0 & 0 & 0 & 0 & -3\lambda \\
0 & 0 & 0 & -2\gamma & 0 & 0 & -\mu & * & 0 & 0 & 0 & 0 \\
0 & 0 & 0 & 0 & -\gamma & 0 & 0 & 0 & * & 0 & 0 & 0 \\
0 & 0 & 0 & 0 & 0 & -\gamma & 0 & 0 & -3\mu & * & 0 & 0 \\
0 & 0 & 0 & 0 & 0 & 0 & -\gamma & 0 & 0 & -2\mu & * & 0 \\
0 & 0 & 0 & 0 & 0 & 0 & 0 & -\gamma & 0 & 0 & -\mu & *
\end{array}
\right).
$$

Here "*" is such that each column sum is zero.

In general the generator matrix is given by the following $(C + 1)(nL + 1) \times (C + 1)(nL + 1)$ tri-diagonal block matrix:

$$
B = \left(
\begin{array}{cccccc}
M_1 & -N & & & & 0 \\
-C\gamma I & M_2 & -N & & & \\
& -(C-1)\gamma I & M_3 & -N & & \\
& & & \ddots & \ddots & \ddots \\
& & & & -2\gamma I & M_C & -N \\
0 & & & & & -\gamma I & M_{C+1}
\end{array}
\right). \tag{9.33}
$$

Here I is the $(nL+1) \times (nL+1)$ identity matrix, M_1 is given by the following matrix

$$
n\lambda I + \left(
\begin{array}{ccccc}
nL\mu & -n\lambda & & & 0 \\
-nL\mu & n\lambda + (nL-1)\mu & -n\lambda & & \\
& \ddots & \ddots & \ddots & \\
& & -2\mu & n\lambda + \mu & -n\lambda \\
0 & & & -\mu & n\lambda
\end{array}
\right), \tag{9.34}
$$

$$M_i = (n\lambda + (C - i + 1)\gamma)I +$$

$$
\begin{pmatrix}
nL\mu & & & & 0 \\
-nL\mu & (nL-1)\mu & & & \\
& & \ddots & \ddots & \\
& & & -2\mu & \mu \\
0 & & & & -\mu & 0
\end{pmatrix}
\tag{9.35}
$$

for $i = 2, \ldots, C + 1$, and

$$
N =
\begin{pmatrix}
n\lambda & n\lambda & & & 0 \\
& 0 & n\lambda & & \\
& & \ddots & \ddots & \\
& & & 0 & n\lambda \\
0 & & & & 0
\end{pmatrix}.
\tag{9.36}
$$

The solution of the steady-state probability distribution of the system has the following meaning:

$$
q_{i(nL+1)+j} = \lim_{t \to \infty} \text{Prob}(x(t) = i, y(t) = j) \quad (i = 0, \ldots, C, j = 0, \ldots, nL).
\tag{9.37}
$$

We remark that the states of the inventory system forms an irreducible Markov chain and therefore the steady-state probability distribution exists. Again there is no analytical solution for $\mathbf{q}$. The BGS method can be used to solve the steady-state probability distribution. We remark that the BGS algorithm converges when applied to solve our problem. One may also employ the CG method to solve the problem with a suitable preconditioner (see Exercise 9.2).

We let

$$
q(i) = \sum_{k=0}^{nL+1} q(i, k), \quad i = 0, \ldots, C
\tag{9.38}
$$

be the marginal steady state probability of the inventory levels of the central warehouse and

$$
t(i) = \sum_{k=0}^{C} q(k, i), \quad i = 0, \ldots, nL
\tag{9.39}
$$

be the marginal steady-state probability of the inventory levels of the the aggregated local warehouses. From (9.38) and (9.39) the expected inventory cost, the expected normal replenishment cost from the plant, the expected emergency delivery cost from the central warehouses, and the expected emergency delivery cost from the plant are given respectively by

$$\begin{cases} I_C \sum_{i=0}^{C} iq(i), \\ R_P \sum_{i=0}^{C} (C-i)q(i), \\ nD_C\lambda(t(0) - q_0), \\ nD_P\lambda q_0. \end{cases} \tag{9.40}$$

9.4 Summary and Discussion

A multi-location inventory system which consists of one depot has been studied. MMPP was used to model the transshipment of demand from the local inventory system to the main depot. The PCG method developed in Chapter 3 can be applied to solving the steady-state probability distribution of the inventory system at the main depot. Under some assumptions on the system parameters, an approximation for steady-state probability distribution was derived when the number of locations is large. The extension of the model to the case of repairable items is possible.

A two-echelon inventory model under the delivery time guarantee policy was then discussed. The system consists of an infinite supply plant, central warehouse, and a number of local warehouses. The problem was considered in three stages. In the first stage, each local warehouse was modeled as a Markovian queuing system. In the second stage, we considered an aggregated model by aggregating the demand and inventory level of all the local warehouses. The results obtained in the first two stages were then integrated. We formulated the problem as an optimization problem.

Exercises

9.1 Suppose a delivery time guarantee policy of guaranteed time T is implemented at each local warehouse in the two-echelon inventory system. We assume that a demand can be served without delay at each local warehouses or by ELT among the local warehouses. This means that whenever there is inventory at any of the local warehouses, a demand can be served within guaranteed time T. The mean delivery times from the plant and the central warehouse to the customers are T_P and T_C respectively and they are exponentially distributed. Find the probability that a demand cannot be served within the guaranteed time T.

9.2 Suppose we take the circulant approximations of M_i and N as follows:

$$c(M_1) = n\lambda I + \begin{pmatrix} n\lambda + nL\mu & -n\lambda & & & -nL\mu \\ -nL\mu & n\lambda + nL\mu & -n\lambda & & \\ & \ddots & \ddots & \ddots & \\ & & -n\mu & n\lambda + nL\mu & -n\lambda \\ -n\lambda & & & -n\mu & n\lambda + nL\mu \end{pmatrix},$$

$$c(M_i) = (n\lambda + (C - i + 1)\gamma)I + \begin{pmatrix} nL\mu & & & -nL\mu \\ -nL\mu & nL\mu & & \\ & \ddots & \ddots & \\ & & -nL\mu & nL\mu \\ 0 & & & -nL\mu & nL\mu \end{pmatrix}$$

for $i = 2, \ldots, C + 1$, and

$$c(N) = \begin{pmatrix} 0 & n\lambda & & & 0 \\ & 0 & n\lambda & & \\ & & \ddots & \ddots & \\ & & & 0 & n\lambda \\ n\lambda & & & 0 & n\lambda \end{pmatrix}.$$

Show that $\text{rank}(B - c(B))$ is independent of L when C and n are fixed where

$$c(B) = \begin{pmatrix} c(M_1) & -c(N) & & & & 0 \\ -C\gamma I & c(M_2) & -c(N) & & & \\ & -(C-1)\gamma I & c(M_3) & -c(N) & & \\ & & \ddots & \ddots & \ddots & \\ & & & -2\gamma I & c(M_C) & -c(N) \\ 0 & & & & -\gamma I & c(M_{C+1}) \end{pmatrix}.$$

Hence construct a preconditioner for large value of L which works well when C and n are fixed.

References

1. Aggarwal P, Moinzadeh K (1994) Order expedition in multi-echelon production/distribution systems, IIE Trans: 26 86–96.
2. Ajmone-Marsan M, Balbo G, Conte G, Donatelli S, Franceschinis G (1995) Modelling with generalized stochastic petri nets, Wiley, New York.
3. Akella R, Kumar P (1986) Optimal control of production rate in a failure prone manufacturing systems, IEEE Trans Autom Control 31: 116–126.
4. Ammar G, Gragg W (1988) Superfast solution of real positive definite Toeplitz systems, SIAM J Matrix Anal Appl 9: 61–76.
5. Axelsson O (1996) Iterative solution methods, Cambridge University Press, Cambridge, UK.
6. Axelsson O, Barker V (1984) Finite element solution of boundary value problems theory and computation, Academic Press, New York.
7. Axsäter S (1990) Modelling emergency lateral transshipments in inventory systems, Manag Sci 36: 1329–1338.
8. Benedetto F, Fiorentino G, Serra S (1993) C.G. preconditioning for Toeplitz matrices, Computers Math Appl 25: 35–45.
9. Berlard P, Saad Y, Stewart W (1992) Numerical methods in Markov chain modeling, Oper Res 40: 1156–1179.
10. Berman A, Plemmons R (1994) Nonnegative matrices in mathematical sciences, SIAM, Philadelphia.
11. Bielecki T, Kumar P (1988) Optimality of zero-inventory policies for unreliable manufacturing Systems, Oper Res 36: 532–541.
12. Buchholz P (1994) A class of hierarchical queueing networks and their analysis, Queue Syst 15: 59–80.
13. Buchholz P (1995) Hierarchical Markovian models: symmetries and aggregation, Perform Eval 22: 93–110.
14. Buchholz P (1995) Equivalence relations for stochastic automata networks. Computations of Markov chains: Proceedings of the 2nd international workshop On numerical solutions of Markov chains. Kluwer, 197–216.
15. Bunch J (1985) Stability of methods for solving Toeplitz systems of equations, SIAM J Sci Statist Comput 6: 349–364.
16. Buss A, Lawrence S, Kropp D (1994) Volume and capacity interaction in facility design, IIE Trans Des Manuf 26: 36–49.
17. Buzacott J, Shanthikumar J (1993) Stochastic models of manufacturing systems, Prentice-Hall International Editions, Englewood Cliffs, NJ.
18. Buzacott J, Yao D (1986) Flexible manufacturing systems: a review of analytic models, Manage Sci 32: 890–905.
19. Chan R (1987) Iterative methods for overflow queueing models I, Numer Math 51: 143–180.
20. Chan R (1988) Iterative methods for overflow queueing models II, Numer Math 54: 57–78.

21. Chan R (1989) Circulant preconditioners for Hermitian Toeplitz systems, SIAM J Matrix Anal Appl 10: 542–550.
22. R. Chan (1991) Toeplitz preconditioners for Toeplitz systems with nonnegative generating functions, IMA J Numer Anal 11: 333–345.
23. Chan R (1993) Iterative methods for queueing networks with irregular state-spaces, Proceedings of the IMA Workshop on linear algebra, Markov chains and queueing models, Springer-Verlag, London, 89–110.
24. Chan R, Chan T (1992) Circulant preconditioners for elliptic problems, J Numer Lin Algebra Appl 1: 77–101.
25. Chan R, Ching W (1996) Toeplitz-circulant preconditioners for Toeplitz matrix and its application in queueing networks with batch arrivals, SIAM J Sci Comput 17: 162–172.
26. Chan R, Ching W (1999) A direct method for stochastic automata networks. Proceedings of the symposium on applied probability, Chinese University of Hong Kong, China.
27. Chan R, Ching W (2000) Circulant preconditioners for stochastic automata networks, Numer Math 87: 35–57.
28. Chan R, Ng K (1996) Conjugate gradient methods for Toeplitz systems, SIAM Rev 38: 427–482.
29. Chan R, Strang G. (1989) Toeplitz equations by conjugate gradient with circulant preconditioner, SIAM J Sci Stat Comput 10: 104–119.
30. Chan R, Nagy J, Plemmons R (1993) FFT-based preconditioners for Toeplit-block least square problems, SIAM J Numer Anal 30: 1740–1768.
31. Chan R, Tang P (1994) Fast band-Toeplitz preconditioners for Hermitian Toeplitz Systems, SIAM J Sci Statist Comput 15: 164–171.
32. Chan R, Ching W, Wong C (1996) Optimal trigonometric preconditioners for elliptic problems and queueing problems, SEA Bull Math 3: 117–124.
33. Chan R, Yeung M (1993) Circulant preconditioners for complex Toeplitz matrices, SIAM J Numer Anal 30: 1193–1207.
34. Chan T (1988) An optimal circulant preconditioner for Toeplitz systems, SIAM J Sci Statist Comput 9: 767–771.
35. Ching W (1997) Circulant preconditioners for failure prone manufacturing systems, Lin Alg Appl 266: 161–180.
36. Ching W (1997) An inventory model for manufacturing systems with delivery time guarantees, Computers Oper Res 25: 367–377.
37. Ching W (1997) Preconditioned conjugate gradient methods for manufacturing systems: The 8th SIAM conference on parallel processing for scientific computing.
38. Ching W (1997) A model for two-stage manufacturing systems: The 5th Mediterranean IEEE conference on control and systems, Cyprus.
39. Ching W (1997) Markov modulated Poisson processes and production planning in manufacturing systems: The WMC'97 international symposium on manufacturing systems, Auckland, New Zealand.
40. Ching W (1997) Markov modulated Poisson processes for multi-location inventory problems, Int J Prod Econ 53: 217–223.
41. Ching W (1998) Iterative methods for manufacturing systems of two stations in tandem, Appl Math Lett 11: 7–12.
42. Ching W (1998) A new model for multi-location inventory problems, Comput Ind Eng 35: 149–152.
43. Ching W (1998) Circulant preconditioners for stochastic automata networks: The 5th Copper mountain conference on iterative methods, Copper Mountain, Colorado, USA.

44. Ching W (1998) Optimal (S,s) policies for manufacturing systems of unreliable machines in tandem: The international symposium on product quality and integrity, Anaheim, USA, 365–370.
45. Ching W (1999) Iterative methods for manufacturing systems: Proceedings of the 2nd world manufacturing congress, Durham, UK, 346–350.
46. Ching W (2000) A model for FMS of unreliable machines: Proceedings of the 4th world CSCC conference, Vouliagmeni, Greece.
47. Ching W (2000) Circulant preconditioning for unreliable manufacturing systems with batch arrivals, Int Appl Math 4: 11–21.
48. Ching W (2001) Machine repairing models for production systems, to appear in Int J Prod Econ.
49. Ching W (2001) Markovian approximation for manufacturing systems of unreliable machines in tandem, to appear in Int J Naval Res Logist.
50. Ching W, Chan R, Zhou X (1997) Circulant preconditioners for Markov modulated Poisson processes and their applications to manufacturing systems, SIAM J Matrix Anal Appl 18: 464–481.
51. Ching W, Chan R, Zhou X (1998) Conjugate gradient methods for Markov modulated Poisson processes: The 2nd Asian mathematical conference, Thailand, 407–416.
52. Ching W, Zhou X (1995) Hedging point production planning for failure prone manufacturing systems: Proceedings of the 5th conference of the operational research society of Hong Kong, 163–172.
53. Ching W, Zhou X (1996) Matrix methods for production planning in failure prone manufacturing systems, Lecture Notes in Control and Information Sciences, Springer-Verlag, London, vol 214, 2–30.
54. Ching W, Zhou X (1996) Machine repairing models for manufacturing systems: The IEEE 5th conference on emerging technologies and factory automation.
55. Ching W, Zhou X (1997) Optimal (S,s) production policies with delivery time guarantee, Lectures in Applied Mathematics, Mathematics of Stochastic Manufacturing Systems The American Mathematical Society, vol 33, 71–81.
56. Ching W, Zhou X (1997) Optimal (S,s) policies for manufacturing systems with buffers holding costs: The 15th world congress of scientific computation modeling and applied mathematics, Berlin, Germany.
57. Ching W, Zhou X (2000) Circulant approximation for preconditioning in stochastic automata networks, Comput Math Appl 39: 147–160.
58. Cho D, Parlar M (1991) A survey of maintenance models for multi-unit systems, Eur J Oper Res 51: 1–23.
59. Concus P, Golub G, Meurant G (1985) Block preconditioning for conjugate gradient method, SIAM J Statist Comput 6: 220–252.
60. Concus P, Meurant G (1986) On computing INV block preconditionings for conjugate gradient method, BIT 26: 493–504.
61. Conway J. (1973) Functions of one complex variable, Springer-Verlag, Berlin.
62. Davis P (1979) Circulant matrices, John Wiley and Sons, New York.
63. Dewan S, Mendelson H (1990) User delay costs and internal pricing for a service facility, Manage Sci 36: 1502–1517.
64. Donohue K (1994) The economics of capacity and marketing measure in a simple manufacturing environment, Prod Oper Manage 3: 78–99.
65. Eben-Chaime M (1995) The queueing theory machine interference model: use and application, Prod Planning Control 6: 39–44.
66. Feller W (1957) An introduction to probability theory and its applications, vol 1, Wiley, New York.
67. Flood J (1995) Telecommunication switching traffic and networks, Prentice-Hall, New York.

68. Flynn B, Sakakibara S, Schroeder R (1995) Relationship between JIT and TQM: practice and performance, Acad Manage J 38: 1325–1360.
69. Freeland J (1980) Coordination strategies for production and marketing in a functionally decentralized firm, IIE Trans 12: 126–132.
70. Golub G, van Loan C (1983) Matrix computations, John Hopkins University Press, Baltimore, MD.
71. Grenander U, Szegö G (1984) Toeplitz forms and their applications, 2nd ed, Chelsea Pub. Co., New York.
72. Heffes H, Lucantoni D (1986) A Markov modulated characterization of packetized voice and data traffic and related statistical multiplexer performance, IEEE J Select Areas Commun 4: 856–868.
73. Hestenes M, Stiefel E (1952) Methods of conjugate gradients for solving linear systems, J Res Nat Bur Stand 49: 490–536.
74. Hill A, Khosla I (1992) Models for optimal lead time reduction, Prod Oper Manage 1: 185–197.
75. Hogg R, Craig A (1978) Introduction to mathematical statistics, Collier-Macmillan.
76. Horn R, Johnson C (1985) Matrix analysis, Cambridge University Press.
77. Hu J (1995) Production rate control for failure prone production systems with no backlog permitted, IEEE Trans Autom Control 40: 291–295.
78. Hu J, Vakili P, Yu G (1994) Optimality of hedging point policies in the production control of failure prone manufacturing systems, IEEE Trans Autom Control 39: 1875–1880.
79. Hu J, Xiang D (1993) The queueing equivalence to optimal control of a manufacturing system with failures, IEEE Trans Autom Control 38: 499–502.
80. Hu J, Xiang D (1994) Structure properties of optimum production controllers in failure prone manufacturing systems, IEEE Trans Autom Control 39: 640–642.
81. Huckle T (1992) Circulant and skew-circulant matrices for solving Toeplitz matrix problems, SIAM J Matrix Anal Appl 13: 767–777.
82. Karmarkar U (1994) A robust forecasting techniques for inventory and lead time Management, J Oper Manage 12: 45–54.
83. Kaufman L (1982) Matrix methods for queueing problems, SIAM J Sci Statist Comput 4: 525–552.
84. Kelley C (1995) Iterative methods for linear and non-linear equations, SIAM, Philadelphia.
85. Kochel P (1996) On queueing models for some multi-location problems, Int J Prod Econ 45: 429–433.
86. Law A, Kelton W (1991) Simulation modeling and analysis, McGraw-Hill, 2nd Edn. New York.
87. Lee H (1987) A multi-echelon inventory model for repairable items with emergency lateral transshipments, Manage Sci 33: 1302–1316.
88. Li L (1992) The role of inventory in delivery-Time completion, Manage Sci 38: 182–197.
89. Manteuffel T (1980) An incomplete factorization techniques for positive definite linear systems, Math Comp 34: 473–497.
90. Meier-Hellstern K (1989) The analysis of a queue arising in overflow models, IEEE Trans Commun 37: 367–372.
91. Moinzadeh K, Schmidt C (1991) An (S-1,S) inventory system with emergency orders, Oper Res 39: 308–321.
92. Monden Y (1983) Toyota production system, Industrial Engineering Manufacturing Press, Atlanta, GA.
93. Nelson B (1995) Stochastic modeling analysis and simulation, McGraw-Hill, New York.

94. Oda T (1991) Moment analysis for traffic associated with Markovian queueing systems, IEEE Trans Commun 30: 737–745.
95. Pidd M (1992) Computer simulation in management science, John Wiley and Sons, 3rd Edn. Chichester.
96. Plateau B, Atif K (1991) Stochastic automata network for modeling parallel systems, IEEE Trans Software Eng 12: 370–389.
97. Pyke C (1990) Priority repair and dispatch policies for repairable item logistic systems, Naval Res Logist 37: 1–30.
98. Ross S (1970) Applied probability models with optimization applications, Holden Day, San Francisco, Cal.
99. Ross S (1983) Stochastic processes, Wiley, New York.
100. Ross S (1985) Introduction to probability models, Wiley, New York.
101. Saad Y (1996) Iterative methods for sparse linear systems, PWS, Publishing Co., Boston.
102. Seila A (1990) Multivariate estimation of conditional performance measure in regenerative simulation, Am J Math Manage Sci 10: 17–45.
103. Sethi S, Yan H, Zhang Q, Zhou X (1993) Feedback production planning in a stochastic two-machine flowshop: asymptotic analysis and computational results, Int J Prod Econ 30-31: 79–93.
104. Sethi S, Zhang Q, Zhou X (1992) Hierarchical controls in stochastic manufacturing systems with machines in tandem, Stoch Stoch Rep 41: 89–118.
105. Sethi S, Zhou X (1994) Dynamic stochastic job shops and hierarchical production planning, IEEE Trans Autom Control 39: 2061-2076.
106. Siha S. (1996) Modeling the blocking phenomenon in JIT Environment: an alternative scenario, Comput Eng 30: 61–75.
107. Sonneveld P (1989) CGS, a fast lanczos-type solver for non-symmetric linear systems, SIAM J Sci Comput 10: 36–52.
108. Stewart W, Atif K, Plateau B (1995) The numerical solution of stochastic automata networks, Eur J Oper Res 86: 503–525.
109. Strang G (1986) A proposal for Toeplitz matrix calculations, Stud Appl Math 74: 171–176.
110. Suri R (1983) Robustness of queueing network formulas, J Assoc Comput Mach 30: 564–594.
111. Tyrtyshnikov E (1992) Optimal and super-optimal circulant preconditioners, SIAM J Matrix Anal Appl 13: 459–473.
112. Van der Vorst H (1982) Preconditioning by incomplete decomposition, Ph. D Thesis, Rijksuniver-siteit te Utrecht, Netherlands.
113. Varga R (1963) Matrix iterative analysis, Prentice-Hall, Englewood, NJ.
114. Yamazaki G, Kawashima T, Sakasegawa H (1985) Reversibility of tandem blocking queueing systems, Manage Sci 31: 78–83.
115. Yan H, Zhou X, Yin G (1994) Finding optimal number of kanbans in a manufacturing system Via stochastic approximation and perturbation analysis, Lecture Notes in Control and Information Sciences, Cohen G, Quadrat J (Eds), Springer-Verlag, London, 572-578.
116. Young H, Byung B, Chong K (1992) Performance analysis of leaky-bucket bandwidth enforcement strategy for bursty traffics in an ATM network, Comput Net ISDN Syst 25: 295–303.

Index